Hot Science is a series explori
and technology. With topics
dark matter to gene editing, these are books for popular
science readers who like to go that little bit deeper ...

AVAILABLE NOW AND COMING SOON:

Destination Mars:
The Story of Our Quest to Conquer the Red Planet

Big Data:
How the Information Revolution
is Transforming Our Lives

Gravitational Waves:
How Einstein's Spacetime Ripples Reveal the Secrets
of the Universe

The Graphene Revolution:
The Weird Science of the Ultrathin

CERN and the Higgs Boson:
The Global Quest for the Building Blocks of Reality

Cosmic Impact:
Understanding the Threat to Earth from Asteroids
and Comets

Artificial Intelligence:
Modern Magic or Dangerous Future?

Astrobiology:
The Search for Life Elsewhere in the Universe

Dark Matter & Dark Energy:
The Hidden 95% of the Universe

Outbreaks & Epidemics:
Battling Infection From Measles to Coronavirus

Rewilding:
The Radical New Science of Ecological Recovery

Hacking the Code of Life:
How Gene Editing Will Rewrite Our Futures

Origins of the Universe:
The Cosmic Microwave Background
and the Search for Quantum Gravity

Behavioural Economics:
Psychology, Neuroscience,
and the Human Side of Economics

Quantum Computing:
The Transformative Technology of the Qubit Revolution

Hot Science series editor: Brian Clegg

QUANTUM COMPUTING

QUANTUM COMPUTING

The Transformative Technology of the Qubit Revolution

BRIAN CLEGG

Published in the UK and USA in 2021
by Icon Books Ltd, Omnibus Business Centre,
39–41 North Road, London N7 9DP
email: info@iconbooks.com
www.iconbooks.com

Sold in the UK, Europe and Asia
by Faber & Faber Ltd, Bloomsbury House,
74–77 Great Russell Street,
London WC1B 3DA or their agents

Distributed in the UK, Europe and Asia
by Grantham Book Services,
Trent Road, Grantham NG31 7XQ

Distributed in the USA
by Publishers Group West,
1700 Fourth Street, Berkeley, CA 94710

Distributed in Australia and New Zealand
by Allen & Unwin Pty Ltd,
PO Box 8500, 83 Alexander Street,
Crows Nest, NSW 2065

Distributed in South Africa
by Jonathan Ball, Office B4, The District,
41 Sir Lowry Road, Woodstock 7925

Distributed in India by Penguin Books India,
7th Floor, Infinity Tower – C, DLF Cyber City,
Gurgaon 122002, Haryana

Distributed in Canada by Publishers Group Canada,
76 Stafford Street, Unit 300
Toronto, Ontario M6J 2S1

ISBN: 978-178578-707-2

Typeset in Iowan by Marie Doherty

Printed and bound in Great Britain
by Clays Ltd, Elcograf S.p.A.

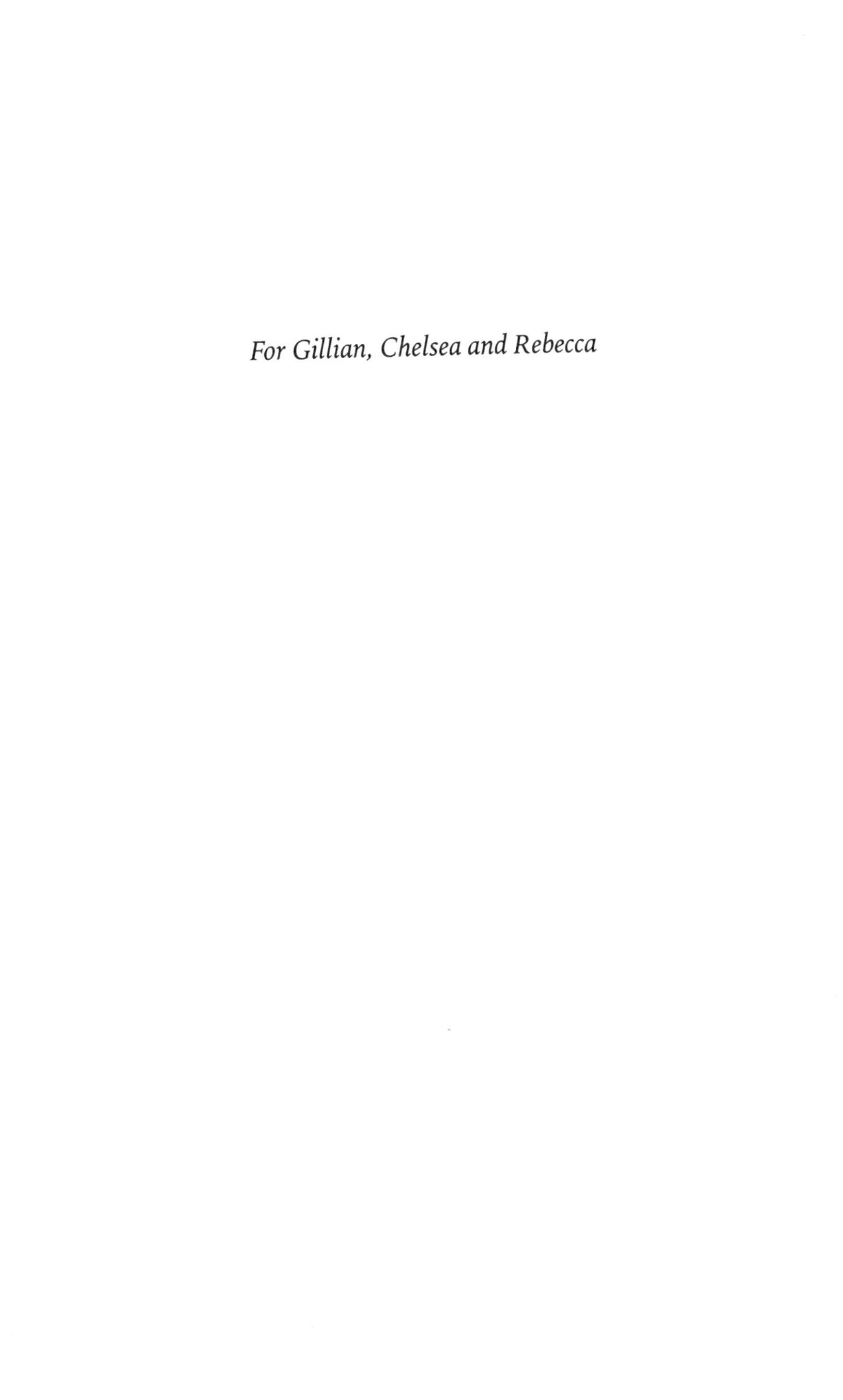

For Gillian, Chelsea and Rebecca

ABOUT THE AUTHOR

Brian Clegg is the author of many books, including most recently *What Do You Think You Are?* (2020) and *Dark Matter and Dark Energy* in the Hot Science series (2019). His *Dice World* and *A Brief History of Infinity* were both longlisted for the Royal Society Prize for Science Books. Brian has written for numerous publications including *The Wall Street Journal, Nature, BBC Science Focus, Physics World, The Times, The Observer, Good Housekeeping* and *Playboy*. He is the editor of popularscience.co.uk and blogs at brianclegg.blogspot.com.

www.brianclegg.net

ACKNOWLEDGEMENTS

My thanks to the team at Icon Books who have helped shape this series, notably Duncan Heath, Robert Sharman and Andrew Furlow.

My interest in computing was shaped by my first exposure at the Manchester Grammar School, where a young teacher encouraged us to punch holes in cards (by hand), send them off to a computer facility in London by post and wait a week for the result to come back in the post. After mixed experiences at university, I fell in love with computing again at British Airways, under the guidance of two mentors, John Carney and Keith Rapley, sadly both no longer with us. It was there that I learned that computer programming is an amalgam of fun and frustration, combining as it does the challenges of puzzle solving and of writing.

Although my coding experience is long in the past, it helps me appreciate the ingenuity of those who attempt to solve the many problems facing anyone who wishes to harness quantum capabilities to bring in a new computer revolution.

CONTENTS

quantum (ˈkwɒntəm)
A minimum amount of a physical quantity that can exist due to the physical limits in nature, meaning that the item can only be varied in such units. Describes the properties of the particles that make up light and matter, which behave entirely differently from familiar objects, with probability at the heart of their behaviour. *From the post-classical Latin 'quantum' meaning amount, quantity or determination of quantity.*

computing (kəmˈpjuːtɪŋ)
The action or an example of calculation or counting. Since the 20th century, the use of computers, particularly electronic computers, to perform computations mechanically. *From the Latin 'computare': to calculate, account, reckon or count up.*

quantum computing (ˈkwɒntəm kəmˈpjuːtɪŋ)
Performing calculations with a device that makes use of the special properties of quantum particles, such as photons of light or electrons, in order to perform certain operations exponentially faster than is possible with a conventional computer.

INSTRUCTIONS FOR A GHOST ENGINE

1

Program one: 1843

In 1840, the British inventor and polymath Charles Babbage gave a number of lectures in Turin, taking as his subject an as-yet-unconstructed device, the Analytical Engine. Born in 1791, Babbage had a sufficiently large inheritance from his father – a goldsmith and banker – never to need to take gainful employment. He enjoyed the social life of the salon as much as his work. It is said that his inspiration to explore mechanical means of calculation was helping out his friend, the astronomer John Herschel, to check astronomical tables. The experience was tedious beyond measure and Babbage is said to have cried out, 'I wish to God these calculations had been executed by steam!'

Had the Analytical Engine ever been built, it would arguably have been the world's first computer in the modern sense of the word. Babbage had been playing the role of a computer for Herschel – that is, a person who undertook calculations. The terminology dates back at least to the seventeenth century – it was only in the mid-twentieth century that the term was shifted from human beings to machines. Although it was entirely mechanical, the Analytical Engine was intended to hold both its data and its programs* on punched cards, based on the cards that had been devised to produce intricate patterns on the Jacquard silk-weaving loom. Unlike its semi-constructed (but never finished by Babbage) predecessor, the Difference Engine, where the instructions on what to do with the data were built into the machinery, the Analytical Engine's instructions could be varied to taste.

Two years later, in an impressive piece of internationalism, a minor Italian military engineer Luigi Federico Menabrea (later to unexpectedly become Prime Minister of Italy) published a write-up of Babbage's Turin talks, written in French for a Swiss publication, the *Bibliothèque Universelle de Genève*. Left in that periodical, this memoir would no doubt have rapidly disappeared into obscurity. However, in 1843 it was translated into English by Ada King, the Countess of Lovelace. In truth, 'translation' is a distinctly weak term for the resultant document, as King added copious notes that tripled the length of the piece, speculating on the future use of the unbuilt Analytical Engine and describing how it could be programmed for a number of tasks.

It is thanks to this single document that Ada Lovelace, as King is usually known, has gained the reputation of being the world's first computer programmer. There is no doubt that Lovelace succeeded in bringing the potential of the Engine to a wider audience, though the degree to which she was indeed

* For those who insist on UK English spelling, it should be noted that the worldwide standard spelling for a computer program is the American one, not 'programme'.

the first programmer has been disputed. One certainty is that the machine these instructions were intended for was never built – realistically, it could not have been constructed with the mechanical tolerances of the time. And so, strictly speaking, we should say that the document contained algorithms, in the form of tables that reflected the structure of the Engine, rather than computer programs in the modern sense.

Algorithms are structured instructions that could be anything from the sequence of actions required to brew a cup of tea to complex manipulations of data to solve a mathematical problem. They don't require any computer – they can be worked by hand – but can, as was the case here, be structured in such a way that they fit well with a computer's architecture.

Unfortunately, in the entirely desirable urge to provide good female role models from the past, Lovelace's contribution has been exaggerated. Lovelace was the daughter of the poet Lord Byron and Annabella Milbanke. As a child, Lovelace was encouraged by her mother to study mathematics. She is often described as a mathematician, but it would be accurate to describe her as an undergraduate-level maths student. From letters between Babbage and Lovelace, it seems well-established that the notes that Lovelace added to Menabrea's work were strongly influenced by Babbage. And even taking algorithms as programs, we know that Lovelace was not the first. This is because, as historian of science Thony Christie points out:

> The Menabrea Memoir that Ada had translated already contained examples of programs for the Analytical Engine that Babbage had used to illustrate his Turin lectures and

> had actually developed several years before. The notes contain further examples from the same source that Babbage supplied to the authoress. The only new program example developed for the notes was the one to determine the so-called Bernoulli numbers.

We do know that Babbage, in his lectures, described algorithms that could have become programs for the Analytical Engine, had it ever been built. Usually with a totally new piece of technology like this we have to wait for some kind of prototype to be constructed before we can be certain of the device's capabilities. But, remarkably, the Analytical Engine algorithms clearly showed the remarkable power that the machine would be capable of, had it ever been built. Rather than wait for programs to be developed, the Engine could instantly leap into action.

Having algorithms that were ready to go on a technology that was impossible to construct at the time was remarkable. That such a thing should happen twice seems even more surprising. Yet 153 years after the publication of Lovelace's translation and notes, a very similar occurrence would play out. This time, the imagined engines in question would invoke the power of the quantum.

Program two: 1996

By 1996, electronic computers had been part of government and business establishments for decades, and had become relatively commonplace in homes, since personal computers moved from the realm of enthusiasts' toys to commercial products in the 1980s. Unlike the Analytical Engine, electronic computers were too complex to be designed by a single person – and although some early programs were the work of an individual, many were constructed by teams.

Although the underlying concepts behind the electronic computer would continue to be developed, and these devices would continue to become increasingly powerful for decades to come, by the 1990s scientists were already aware of limitations in the way that such computers worked. Fifteen years before our key date, the physicist Richard Feynman had speculated about constructing computers where the basic unit of operation was not the traditional bit, which was limited to holding values of 0 or 1, but rather a 'quantum bit', based on a quantum particle such as a photon or electron, which could be in an intermediate state with a potentially infinite set of possibilities.

By the 1990s, teams were thinking about or attempting to construct such quantum computers. The challenges they faced were huge. In 1996, ability to manipulate individual quantum particles was in its infancy. The technology required to build a quantum computer was entirely impractical. Yet, just like the algorithms for the Analytical Engine, it proved possible to devise an algorithm for quantum computers that, should the machines ever work, could revolutionise the business of searching for data – a task that lies at the very heart of the computer business.

This particular quantum algorithm was devised by Lov Grover, then working at Bell Labs in America. Grover was born in 1961 in the North Indian city of Roorkee. His original intention was to become an electrical engineer – perhaps not a surprising ambition, as he lived in the city that was home to Asia's first specialist technology establishment, founded in the 1840s to give training in engineering. However, Grover did not attend the University of Roorkee (now the Indian Institute of Technology Roorkee), but rather the Indian Institute of Technology in Delhi.

A career in electrical engineering was still Grover's intention when he emigrated to the USA, but by 1985, when he had achieved a PhD in the subject at Stanford University, his interest had grown in the quantum applications of physics – his doctorate concerned a device that typifies the oddities of the quantum world, the laser, and it was quantum physics that drove his thinking when he joined Bell Labs.

Then the research arm of the communication company AT&T, Bell Labs was not unlike the research department of a modern technology giant such as Google or Apple today. Researchers at the laboratories were given an impressive degree of freedom to explore new ideas, as a result of which the company was rewarded with dramatic developments – often in and around computing. It was at Bell Labs, for example, that John Bardeen and Walter Brattain, under the direction of William Shockley, had come up with the transistor. In the 70s, Bell had been responsible for developing UNIX, the operating system that still is central to much computing, and the C programming language which, with its derivates, still dominates the world of conventional computer programming.

When Grover joined Bell, the idea of developing algorithms that could run on quantum computers was already in the air, with the first devised in 1994. The freedom to do what Grover described as 'forward-looking research' was still available at Bell and he almost immediately devised his quantum computing search algorithm. He would later write up his new idea in the journal *Physics Review Letters* as 'Quantum Mechanics Helps in Searching for a Needle in a

Haystack'. His idea could, in theory, transform the business of searching.

At the time, search engines* as we now know them were in their infancy. Prior to this point, if you were an early adopter of the internet and wanted to find your way around the World Wide Web, you would use a curated list of links – a manual index. The leading search engine in 1996, AltaVista, only started operation the year before. Google would begin life as a research project in 1996. But though search engines per se were a novelty, databases had been at the heart of much computing for decades – the need to quickly find and retrieve data was the driving force behind much commercial computer use, and anything that could speed up that process would clearly be attractive. What Grover realised is that with a quantum computer, he could not only speed up searching, he could supercharge it.

We can think of a database as an electronic version of a card index. Each database consists of a set of records, with a record being the equivalent of a single card. To find our way around the records, the database is indexed. In the card index, that is done simply by putting the cards in alphabetical order of the heading – but an electronic database can have multiple indices, enabling it to effectively re-order the cards depending on a whole range of pieces of data on the card. So, for example, you could find a customer by name, or phone number, email, or address, or purchases.

To perform this task, databases have clever get-arounds that enable them to get to information that isn't stored in an ordered way faster than would be the case if they were

* The term 'search engine' is such a part of modern life that we tend to forget that the anachronistic use of 'engine' in this context is an intentional reference back to Babbage's engines.

simply to search through item by item until they reached the correct record. The same goes, on a tremendous scale, for a search engine like Google. It would be ridiculous to expect Google's software to look through the entire content of the web, with well over a billion websites, some of which are enormous in their own right, every time we type in a query. According to Google, their index alone is over 100,000 terabytes in size (in bytes, that's 10^{17} – 1 followed by 17 zeros) and is growing all the time. Searching through unstructured data like the net is a nightmare.

To get a feeling for the problems of dealing with unstructured data, think of an old-fashioned phone book. You've got a list of phone numbers ordered by the name of the people or businesses to whom they correspond. So, given that name, you can rapidly home in on the correct entry and pull out the phone number. But imagine you had the phone number and wanted to find the corresponding name. Making a search the other way round would be a nightmare. You would have to look at each entry in turn and see if you found a match. If, for example, your phone book had 1 million entries, on average you would need to look at 500,000 of these before hitting on the correct entry. If you were ridiculously lucky it could be the first item you checked, but if you were unlucky you might need to check every one of them.

If you are thinking how easy it is to look up a piece of information such as a phone number online, remember that's only because someone has already done the hard work for you and indexed the information, effectively turning it from an unstructured collection of data into the equivalent of a phone book that can be sorted on any part of the entry. But with Grover's algorithm running on a quantum computer, things would be radically different. Instead of potentially looking through the entire list, Grover's algorithm guarantees to get to your result with a maximum number of attempts that is the square root of the number of entries – in the case of our million-entry phone book, that's a limit of 1,000 tries. The algorithm's abilities grow explosively faster than an item-by-item search. And that's just the beginning – as we'll discover later, in 2000 Grover came up with another quantum computing algorithm that makes it possible to do fuzzy searching, which is a whole new ballgame.

It's no surprise that Google is now one of the biggest investors in quantum computing technology. This one algorithm alone could have a huge impact on their business – and it is not the only way that quantum computers could far outperform their traditional equivalents. Although a lot of work on developing quantum computers and quantum algorithms is taking place in universities, companies such as Google and IBM (something of a database specialist) are probably the biggest players.

Before we dive into the workings of quantum algorithms and see how Grover's algorithm can perform this amazing high-speed searching (or, for that matter, how another early quantum algorithm, Shor's factoring algorithm, could render current internet security useless), we need to take a few steps back. Most of us use computers on a daily basis without having a clue about what's going on inside. To provide the necessary context, the next two chapters will give an introduction to the workings of computer hardware and computer algorithms and the associated programs. That, in effect, gives us the toolkit for the traditional computer that you may have on your desk or in your pocket in the guise of a smartphone.

We then need to get up to speed on quantum physics, and specifically how a quantum computer operates differently from a conventional electronic computer (which is a quantum device in its own right). Finally, we can see just what quantum computing algorithms are capable of, why it has taken so long to get to a workable quantum computer and what it may be able to do for us in the future.

Let's get started, then, where Charles Babbage and Ada Lovelace left off – with the invention of the computer itself.

2 MAKING A WORLD BIT BY BIT

It might seem entirely reasonable that the idea of Babbage's Analytical Engine, combined with the programming possibilities described in the Menabrea memoir, would be sufficient to inspire others to construct computers. However, the usual picture of Babbage as the 'father' of computing is just as wildly over-romanticised as making Lovelace the first programmer. In reality, the Analytical Engine was a novelty concept, unachievable in practice, that had no more impact on real life than the heat rays in H.G. Wells' novel *The War of the Worlds* did on the development of weaponry.

If anything, the next step along the road to modern computers involved taking a step back when the American inventor Herman Hollerith provided technology to help with the 1890 US census. As a result of the growth of population and far more information being collected than in the first censuses, it was taking longer and longer to process the data that had been collected. It took administrators a whole eight years to collect and tabulate the data from the 1880 census – the fear was that, before long, it would take longer to process the data than the ten-year gap between censuses being taken.

Like Babbage's theoretical engine, Hollerith's equipment made use of punched cards – rectangular cards with a grid of numbered positions on them, though Hollerith's held significantly more data than Babbage had envisaged. Hollerith's earliest cards were unprinted, but they soon had a grid of numbers to make it easier for humans to check them. The first cards were divided into twelve rows of 24 columns. Over time, the number of columns packed in was increased, in part by changing from circular holes to narrow rectangular ones, until they reached a standard of 80 columns. Although the size wasn't initially consistent, the cards would end up $3\frac{1}{4}$ inches by $7\frac{3}{8}$ inches, apparently based on the size of banknotes.

Holes were punched in some of the available positions to record the data. Eventually, sophisticated typewriter-like card-punching machines were developed, though initially much of the punching was done by hand. Such cards would be very familiar to anyone who worked on computers in the 1960s and early 70s, when information was still entered into computers using a deck that could potentially contain

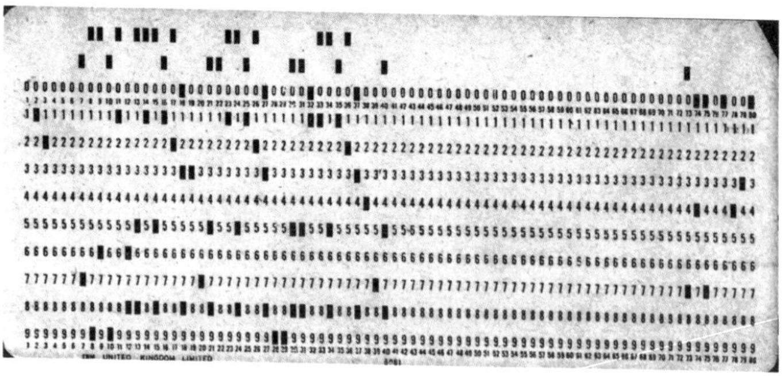

An IBM punched card from the 1960s.

thousands of cards. A line on the card was referred to as a 'Hollerith string' in the inventor's honour.

There was a fundamental difference between Hollerith's devices and the kind of computer that used the cards later on. Hollerith's machines provided only two functions – counting and sorting cards, dependent on the position of the punched holes. Unlike the Analytical Engine or those twentieth-century computers, the algorithm – the logic of *what* to do – was fixed in the structure of the machines, hence the huge step back from the promise of the Analytical Engine. Hollerith's cards could be used, for example, to count the number of children of a certain age in a city, or to provide a list of names in alphabetical order, but there was none of the flexibility that we (and Babbage) would expect.

However, at the time, the relative simplicity was arguably something of a benefit. The cards and machines proved to be a huge success. The census was saved, and soon this data-processing revolution would be available to a whole range of businesses, government departments and universities. Hollerith's Tabulating Machine Company later joined with three other companies to become the rather better-known International Business Machines, which morphed into IBM. But true computers, in the sense implied by the Analytical Engine, would not become a reality until the end of the Second World War.

Dr Turing's universal machine

Mechanical computing could only go so far. Even though it became more practical to engineer the extremely fine gears to the tight tolerances required for Babbage's design, and the Hollerith machines gradually added new ways to sort and shuffle the cards, it would take a more nimble approach to get to the kind of flexibility Babbage had envisaged. Mechanical calculators would continue to be used into the 1960s, but it was the switch to electronics which enabled the truly flexible computer to move from dream to reality.

There is considerable dispute over where the first truly programmable electronic computer was constructed. Colossus, built for the Bletchley Park codebreakers in the UK in 1943, and ENIAC, developed for artillery calculations in the US in 1945, both have reasonable claims to the label as a result of technical differences between the two. The important step forward was that where mechanical predecessors were limited by the physical difficulties of getting gears (or later electromechanical switches) to perform sufficient operations in a small enough space, Colossus and ENIAC made use of the abilities of electronic devices where the moving parts were electrons, enabling impressive miniaturisation.

That's not to say that these early electronic computers were compact – quite the reverse. Despite having a tiny fraction of the capabilities of the most basic phone, they typically filled a large room and depended on electronic devices called valves (known as vacuum tubes in the US), which switched or amplified electrical signals to perform the basic functions that would be required to perform computations. However, much of the theory they depended on had been developed before these electronic devices were created, by two outstanding individuals – Alan Turing in the UK and John von Neumann in the US.

Born in London in 1912, Turing was educated at Cambridge, gaining a doctorate at Princeton before returning to the UK. Soon after getting back to Cambridge he would join the British cipher-cracking centre at Bletchley Park. For a long time, Turing was a hidden figure in British history. This was the result of a combination of security concerns from his involvement with Bletchley Park, the very existence of which was kept secret long after the war, and the British establishment's reaction to Turing's homosexuality. Turing was a wartime hero, now recognised as one of the greatest individuals of the twentieth century, and yet he was treated appallingly by the state, receiving a prison sentence and chemical treatment for his 'crime' of homosexuality.*

It is often said that Turing committed suicide as a result of the treatment he received. This, however, is pure speculation from a poorly conducted inquest. In fact, all the evidence from the time was that Turing was in good spirits shortly before he died and had recovered from the chemical abuse he received. Turing died at the age of 41 from cyanide poisoning and it was assumed that this was as a result of eating a deliberately poisoned apple, perhaps inspired by seeing the Disney film *Snow White and the Seven Dwarfs*. However, shockingly, the apple, which was found by his bed, was never tested for poison. And, at the time of his death, Turing had been running an experiment in a room adjacent to his bedroom which could have given off hydrogen cyanide fumes. Many now believe that his death was a tragic accident.

Turing would help build the Manchester Baby, the first stored-program electronic computer** in the late 1940s, but his most important contribution to the computing (and hence

* Turing was belatedly pardoned by the UK government in 2013.

** A stored-program computer is one that holds its program in electronic memory, rather than setting up the program using switches or plug boards, as was the case with the very first electronic computers such as Colossus and ENIAC.

quantum computing) story came considerably earlier, in 1936, when he wrote a paper titled 'On Computable Numbers'. In part, the paper was devised to address the *Entscheidungsproblem* (decision problem), a mathematical challenge from the 1920s which asked whether it was possible to have a mechanism that would be capable of deciding whether any mathematical statement was universally valid, given a set of axioms.

Axioms provide the foundations of mathematics. They are the basic assumptions that don't need to be proved, and it's impossible to do maths without them. Some examples are those for Euclidean geometry, such as 'a (straight) line can be drawn from a point to any other point' or those for set theory, which are necessary for all arithmetic to work. The mathematician Alonzo Church devised a mathematical solution to the *Entscheidungsproblem* at around the same time that Turing used a different approach in his paper. Both decided it was not possible to have such a mechanism – but Turing's approach had a wider applicability because the way he looked at the problem involved using an imaginary 'computing machine'.

This imaginary device (it would be impossible to create it practically) was nothing like the kind of machine we now think of as a computer. However, it is known as a universal Turing machine because it is capable of undertaking *any* process that a computer can perform, despite being extremely simple in its structure. Admittedly, it would be painfully slow to carry out even the simplest computing task, but it could get there, given sufficient time.

Turing's imaginary machine consists of three principal components. The first is a limitless piece of 'tape' (he described this as being 'the analogue of paper'), which is divided into sections that Turing called squares, each of which could be blank or could contain a symbol. The second part is a read/write head, a device which can move up and down the tape, one square at a time, either reading a symbol, writing a symbol or erasing a symbol. Finally, there is a controller which contains a set of rules. These rules are not the computer program to be executed, but rather what we would now call the operating system – they instruct the head how to behave depending on what it is has read or written, and tell it how to start and finish a run.

What the machine writes in a square (or reads from it) can either be data in the form of a zero or one, or it can be an instruction code, which could instruct the machine to, say, erase a value, move one to the right, read a value and move left or right depending on that value. This is how the tape can hold both data and the program – essential for a universal computer. Note that there is no distinction on the tape between data and program – the whole thing can be mixed up however we like to achieve the desired outcome. And the tape does not have to begin blank (in fact, very little could be achieved if it did) – the initial data and instructions can already be printed on the tape when a run starts.

With this very small set of possibilities, anything that is computable can be computed. Of course, this isn't what goes on inside your computer (let alone in your phone) – but any of the results achievable via the internal workings of any computer could be generated by a universal Turing machine.*

* A universal Turing machine could not itself take input from a keyboard, or display the outcome of its calculations on a screen. We have to get the information onto the tape first, and something would have to take its results and convert them into visible form. And every real computer will have specific input and output controllers to handle this. But the Turing machine would provide all the core computing functions provided by any physical computer.

This is because the machine is a stored-program computer with unlimited capacity (remember, the tape is as long as we need it to be). The machine can act differently according to the information it reads off the tape. The tape holds data to go into the operation and data then comes out (the result), but crucially it also holds the program – the particular combination of basic actions that will produce the desired outcome. Just like the tables that were used to structure the algorithms for Babbage's Analytical Engine, these programs are algorithms written in a form that is specifically suited to the way that the universal Turing machine works.

Turing's theory lies behind the majority of modern computers, including the Manchester Baby. But theory was not enough to get to working computers like this: the other essential foundation would come from the other side of the Atlantic.

The architect of computing

John von Neumann was born in Budapest in 1903, where he proved to be a mathematical prodigy, said at the age of six to be able to divide one eight-figure number by another in his head. He simultaneously took a chemical engineering degree at the Swiss Federal Institute of Technology (ETH) and a PhD in mathematics at Pázmány Péter University, Budapest. After a few years in European universities he took up the offer of a lifetime professorship at the Institute for Advanced Study in New Jersey. It was from his work here that he would become a leading light in the mathematical side of the Manhattan Project to build a nuclear weapon.

After the Second World War, von Neumann became heavily involved in computing, devising a sorting algorithm for the early EDVAC computer* and coming up with mechanisms for generating pseudo-random numbers. Random numbers are important for many computer simulations of real life and are at the heart of the Monte Carlo method, named after the casino. Von Neumann developed this procedure to help with nuclear research, but it would be used in many different algorithms that try to reflect the real world. For example, when trying to work out how queues build up, a researcher would repeatedly run a program using randomly selected arrival times from a distribution that matched typical occurrences in real life.

Because of its nature, a computer cannot generate truly random numbers – given the exact same starting point it will always produce the same outcome, where randomness by definition has to be unpredictable. But a *pseudo*-random number gives a reasonable approximation to randomness by taking a 'seed' or starting value typically selected from the time of day and running it through a formula that generates very different outcomes with small variations in input.**

Most significant for the computing future, though, was von Neumann's devising of a practical physical architecture that could implement Turing's stored-program concept. Von Neumann based his work on the approach used in the early ENIAC computer, but made it more general and widely applicable. The term 'architecture' here is not exactly the same as the use of the word in normal English. Rather than

---

* EDVAC was the successor to ENIAC. Though still based on vacuum tubes, it had two big enhancements: being a true stored-program computer, and using binary rather than decimal values.

** We'll find out more about the Monte Carlo method and pseudo-random numbers versus actual random numbers on page 85 as they are important in the use of quantum computers.

describing the built environment itself, architecture in computing is more like the architect's drawings: the conceptual structure of the finished item, whether it be hardware, as in this case, or computer software to run on that hardware.

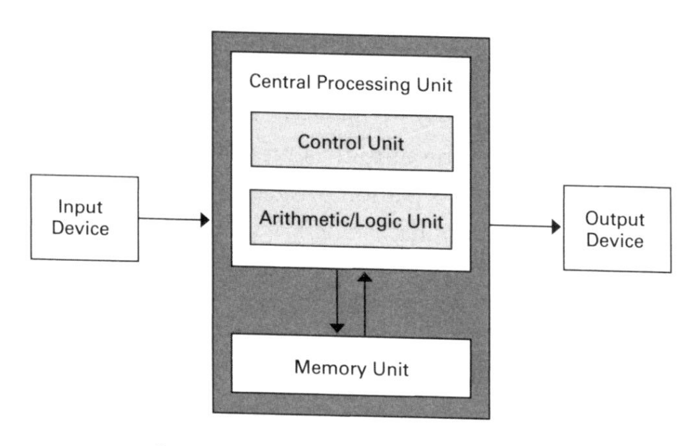

Simple von Neumann architecture.

At the heart of von Neumann's design for computer architecture are two familiar items that come up when choosing a PC or phone: a central processing unit, or CPU, and a memory unit. Between them these correspond to a combination of the tape and the read/write head in Turing's machine. As its name suggests, the memory unit remembers the data and the stored programs, while the CPU manipulates the data based on its built-in instructions (a so-called 'instruction set'), which are triggered by the program. Von Neumann's CPU was split into two, the control unit – which tells the other parts of the computer what to do and handles timings to ensure that operations are carried out in the right order – and the arithmetic/logic unit – which handles basic mathematical functions and logical operations.

To finish off his design, von Neumann incorporated two very broad concepts – the input device, by which data and programs were to be entered into memory, and the output device, where the results produced during the computation were to be presented to the user. Initially the input might have been punched cards or paper tape and the output would be onto more paper tape or as a printout on paper. Now, we're more used to the input being keyboard, mouse and touchscreen, while the output is primarily onto a screen, with only selected information printed onto paper.

All kinds of additional elements would be added to the architecture over time. For example, a floating-point unit, to deal with so-called 'floating point numbers', effectively approximations to a real number with decimal places such as 8.1258727… – or, to handle the specialist number-crunching required to deal with graphics, a graphics processing unit, or GPU. Typically, the CPU hands the specialist calculations off to these separate units, which do the necessary work and then hand the results back to the CPU. As we will see, a similar role is envisaged for quantum computations in the future.

Another important part of the architecture that von Neumann didn't include in his consideration was the 'bus' – effectively the set of wires that allow data to flow from one part of the computer to another, or to transfer data between the outside world and the inner workings of the computer. We are now very familiar with external versions of this such as USB (Universal Serial Bus), but the internal bus that links different modules is even more crucial.

Finally, we now tend to have a much larger extension of memory called storage to keep hold of data outside of the immediate requirements in memory to handle a computation.

Initially, magnetic tapes were frequently used for this (which is why early movies showing computers always tended to have tapes whizzing back and forth). Now, storage is usually provided by magnetic discs or specialised memory that doesn't lose its values when the power is switched off. Such data can be held between runs of the computer, or archived for future access. However, the basics of von Neumann's architecture remain at the physical heart of your laptop, tablet or phone.

Bits and bytes

To get closer to what's different and special about quantum computers we need to zoom in on two specific parts of the hardware – the computer's working memory, and tiny structures within the CPU known as gates. As we've already discovered, memory is rather like Turing's tape – a place where information can be stored, retrieved and modified. There is one subtle difference, though – unlike the squares on Turing's tape, which could contain a complete instruction, each part of the computer memory can only hold a single value of either zero or one. This 'binary' nature is embedded in the familiar word 'bit', which is a contraction of 'binary digit'.

You can think of a memory chip as being like a series of little boxes, each of which can either be empty (0) or have a marker in it (1). In practice, of course, the memory in a computer is not made up of an array of boxes. It's an electronic device, which in a modern computer will be based on so-called MOSFETs located on the silicon wafer of a chip. MOSFETs are Metal Oxide Semiconductor Field Effect Transistors. Each bit is a small circuit, typically a combination of one or more transistors to control the value, and a capacitor – a component that will either hold an electrical charge to represent 1 or have no charge to denote 0.

The bits are then grouped into small collections known as words, each of which has a separate address in the form of a binary number. Just as your postal address locates your specific house or flat on a street, town and country, so the word's address pins down its location in a specific chip. That means that the CPU can go straight to a specific word, rather than having to hunt for it, hence the name 'random access memory' or RAM.* Information storage like a reel of tape or a book has to be accessed sequentially until the correct point is reached, but with random access memory, each word has a specific address and the computer can jump straight to it. Word size (which typically comes in powers of two – so 2, 4, 8, 16, 32, 64...) has got larger over the years. The first personal computers used 8-bit words, but by the 1990s 32-bit words were common, and now the majority of computers use 64-bit words.**

Confusingly (it would be boring if it were too simple), the words also tend to be divided up into 8-bit chunks called bytes. This is for convenience, as when handling text, each byte can hold a single character. An 8-bit byte can have 256 values from binary 00000000 (value 0) to 11111111

---

* This name made sense originally, but doesn't now. You may have come across, for example, ROM or flash memory, which are different from RAM in the mechanism used to store data, but are all still random access memory. Non-random access memory dates back to ancient memory techniques where data might be temporarily held in devices such as mercury delay lines, where the data always had to be accessed in the same sequence.

** To be able to jump directly to a specific bit in a word, the computer's internal bus needs as many wires as there are bits in a word. So, a modern computer might be described as having a '64-bit bus'.

(value 255). Eight bits is a convenient size, as the common standard to represent characters, ASCII (American Standard Code for Information Interchange), requires 7 bits for each character, allowing 128 characters in a character set. So, for example, the ASCII code for the capital letter A is 65, which in binary would be represented as 1000001, each digit being a single bit in the relevant byte. The extra bit was often used for a checking mechanism known as a parity bit, and having 8 bits also fitted with IBM's proprietary 8-bit EBCDIC code used on many mainframe computers.

Having 128 characters (in practice rather fewer than this are available, as over thirty of them were originally used as 'control characters' to manipulate printers or control magnetic tapes and these have not been repurposed) proved distinctly limiting, especially as computers spread around the world and a whole range of languages had to be accommodated. As a result, although ASCII remains at the heart of the encoding system, it was extended in the late 1980s in the form of Unicode, an extended coding approach that can provide 16 or even 32 bits for a single character, which allows us to have the much wider range of characters and symbols we now expect from our computers.

Realistically, this means that the byte is not incredibly meaningful anymore, as it no longer represents a single character and isn't an important size in the physical internal structure of a computer – but it has stuck as the measure used for memory and storage, which at the time of writing typically comes in gigabytes (roughly billions of bytes) or terabytes (roughly, thousands of billions of bytes). By contrast, data transmission speeds are measured more logically in bits – so your download speed from the internet, say, is usually given in megabits (millions of bits) per second.

Gated communities

It's not enough, though, that we have a way to store data. The name of the game is not just storing data, but processing it. The computer's processor has to be able manipulate data by using logical operations. These allow it to do everything from arithmetic to selection and sorting. And where the fundamental unit of memory is the bit (or, at a push, the word), the fundamental unit of processing is a gate.

Gates usually work on one or two bits at a time, performing the most basic of manipulations, which can then be combined in different ways to produce more and more sophisticated operations. You can think of a gate as a black box that takes an input from one or two bits and spits out a resultant value, usually as a single bit, into another piece of memory (or the same one).

We'll take a look at the basic gates, as it is impossible to really understand what quantum computers do without first having an idea of gates, which take on a whole new level of operation in the form of quantum gates. The simplest of all the conventional gates is the NOT gate. This takes the value in a bit and sends out the opposite value. So, if the bit has the value 0, the NOT gate outputs 1. If the bit has the value 1, the NOT gate outputs 0. This may feel trivial, but bear in mind that everything in a digital computer,* whether data or programs, comes in the form of zeros and ones.

---

* It's distinctly odd that we refer to 'digital computers', as ever since the 1940s, they have not worked in the standard *digital* form, which is decimal (because 'digital' implies counting on your fingers, using the numbers between 0 and 9), but instead have used binary (just 0 and 1). The term is used as a contrast with 'analogue', meaning a computer that works on continuously variable quantities, typically because it uses physical processes such as flows of liquids, levels of

It's not necessary to know the detail of how each of the gates we will meet is performed by electronic components, but just to get a feel for the kind of complexity we're dealing with, the simplest NOT gate can be achieved using just one transistor and a couple of resistors, although in practice, two-transistor gates are more common. Whatever the circuit, the result is that if a small signal (voltage) enters the NOT gate a (relatively) large one comes out, and if a large signal enters, a small one comes out.

The other main forms of gate take two inputs and produce one output – so they are able to make comparisons between values. In all the following examples, 0 is used to represent a low voltage signal and 1 a high one. The AND gate produces 1 if both its inputs are 1 – otherwise it produces 0. If we think of it purely in terms of bits, it outputs 1 if, and only if, both input bits have 1 in them. The OR gate, by contrast, is less fussy. It produces 1 if either or both of the inputs are 1, only producing 0 if both inputs are 0.

The next pair of gates are negative versions of AND and OR respectively – NAND and NOR. These produce the exact opposites to their counterparts. So, where AND only produces 1 if both inputs are 1, NAND produces 1 *unless* both inputs are 1. Similarly, NOR produces 1 only if both inputs are 0. You could produce a NAND by applying a NOT gate to the output of an AND gate, and a NOR by putting a NOT gate after an OR gate.

Finally, we get exclusive with the XOR and XNOR gates. (The X is a rather clumsy contraction of 'exclusive'.) As we

electrical currents or the lengths of objects. Strictly, the opposite of analogue is quantum, but we don't use this as 'quantum' has come to refer specifically to the physics of quantum particles. More on this in the next chapter but one.

have seen, the OR gate produces 1 if *either or both* inputs are 1. But XOR produces 1 if *either but not both* inputs is 1. Similarly, XNOR produces 1 either if both of the inputs are 0 or if both inputs are 1.

In practice, two of the gates are super gates that can reproduce the effects of the others. Either NOR gates or NAND gates in combination can produce all the other effects. Using NOR gates, for example, the simplest of the combination gates, as might be expected, is the NOT gate, which can be produced by sending the same input to both inputs of the NOR. An OR gate can be made by sending the output of NOR into both inputs of a NOR (effectively making a NNOR), and so on. Using such universal gates is less efficient, as it will always take more transistors to make any particular compound gate, but it still may be done to simplify design.

Matching and arithmetic

Once we have the ability to use gates, it's possible to use them to start interrogating memory, making comparisons and performing arithmetic. You could imagine, for example, a very simplistic equivalent of Hollerith's punched cards where, for example, each bit identified a different aspect of a car in a car showroom. For example, you could have bits representing each colour option, each model and so on. (This wouldn't be done in reality, as it's a very inefficient way to represent the data, but it gives a feel for logic circuits in action.)

To select out, say, all the red cars you would simply need to use an AND gate to combine the values in a piece of memory which has just the red bit set to 1 with the chunks of memory containing the car details, one car after another. If the answer comes back with a 1 in that bit it's a red car, if it comes back 0 it's not a red car.

Doing arithmetic is a bit more complicated, but in principle it only involves two logical operations: AND plus XOR. To see how this works, we need to take a step back and look at binary arithmetic – as that is all our computer will ever be working with, even though it doesn't come naturally to us with our focus on decimal values.

Let's take the simplest possibility – adding the contents of two bits. Each bit can only have the value 0 or 1. The results are as follows:

$$\begin{array}{r} 0+ \\ 0 \\ \hline 0 \end{array} \qquad \begin{array}{r} 0+ \\ 1 \\ \hline 1 \end{array} \qquad \begin{array}{r} 1+ \\ 0 \\ \hline 1 \end{array} \qquad \begin{array}{r} 1+ \\ 1 \\ \hline 10 \end{array}$$

The first three make sense to brains that are brought up on decimal arithmetic. The last one is a little more of a leap, but remember that in binary each column of a sum can only hold 0 or 1. In familiar decimal arithmetic, when you hit the maximum (9) and add one more you reset that column to 0 and carry 1 forward to the next column.

$$\begin{array}{r} 9+ \\ 1 \\ \hline 10 \end{array}$$

Similarly, in binary, when you hit the maximum value (1) and add one more, you reset that column to 0 and carry 1 forward to the next column. This is why allegedly witty

T-shirts carry the slogan 'There are 10 types of people in the world: those who understand binary, and those who don't'.

If we treat the two values going into our addition as the inputs to a gate, the value in the rightmost column of the answer is the result of applying XOR to the two inputs, while the amount carried forward one column to the left in the answer is the result of applying AND to the two inputs. With just these two logic gates, we have got the basics of a device to perform addition.

It's not all there, of course. As it stands, our adder can only combine two bits: we would require a large number of these circuits, one for each column of a long number, to make our device useful. And, technically, what we have here is what's known in the trade as a 'half adder' – it doesn't do the whole job. It is able to carry a value forward to the next column to the left, as in the example above, but it doesn't know what to do if the column to the right feeds a carried forward value into it. If we imagine having such an adding device operating on the second column from the right of a larger number, it doesn't take into account the possibility of a carry-over from the rightmost column, which needs to be included to make a 'full adder' – but this is relatively easily done with a few more gates.

Similarly, full subtraction can be achieved using XOR, AND, OR and NOT gates. Once you have the mechanism for adding, subtracting and carrying, moving on to multiplication and division is relatively simple, as the processes can be built from repeated application of the simple operations. So, for example, we can multiply number A by number B by repeatedly adding A to a total B times. Combine these different arithmetical abilities and you have the core of the mathematical operations that enables a computer to do far higher-level working.

Scaling up

In a real-world computer, of course, you don't see individual gates, because they are just tiny, invisibly small parts of a huge chip containing many billions of transistors. In principle, though, you could build a very simple computer where each gate was a separate physical object. When I was about ten, I had a mechanical computer that did just this. Called the Digi-Comp 1, it had a series of plastic sliders which were moved using a manual pull tab. The sliders were linked by mechanical gates that combined metal wires and plastic pegs and could be used to add and subtract three binary columns just as was done with two columns in the example above.

Similarly, the maths communicator Matt Parker produced a mechanical computer that could count up to sixteen by using large numbers of falling dominoes to create the appropriate gates. It's pretty well impossible to describe the domino computer, but there is a video of it in operation, available online.* Of course, if you were to open up your computer or prise apart your phone (not recommended), you would see instead a large microchip providing all the functions of the CPU and one or more other chips that hold the memory.** Within these bland-looking plastic rectangles

---

* Visit https://tinyurl.com/DominoComp to see the domino computer in action.

** There will also be some memory on the main processor chip for immediate access.

is a phenomenally complex integrated circuit, etched on a thin sliver of silicon.

The computer I am writing these words on has a quad core i5 processor – by no means the most powerful available – which features around 2 billion transistors built into a single chip. A typical gate uses between three and five transistors – so that provides around 500 million gates. Enough for a whole lot of processing. Similarly, my computer has 24 gigabytes of memory: 206,158,430,208* bits. The scale is hard to get your mind around, but the way these hardware parts function (admittedly with lots of added bells and whistles) is merely a pumped-up version of the simple examples above.

We might now have an idea of what's physically under the hood, but the hardware alone is useless. A phone or computer without software is just an expensive paperweight. We need to add in the instructions that will bring the computer to life, and that means starting by getting a feel for the nature of algorithms.

---

* Pedantically, a kilobyte of memory is not 1,000 bytes but 1,024 bytes (to easily work with binary, it's made a power of two), and so it continues up the scale, so 24 gigabytes is significantly more than 24,000,000,000 bytes, which would only be 192,000,000,000 bits. Confusingly, disk storage is measured using decimal rather than binary values, so a 500-gigabyte hard disk really does have 500,000,000,000 bytes (though not all of it is accessible). But because modern SSD drives are actually memory, they use the binary version, making a 500-gigabyte SSD bigger than a 500-gigabyte hard disk. There is technically a unit specifically for the larger binary version, the gibibyte – but no one uses it.

3 THE SOFT TOUCH

Insert a teabag into a cup. Add fresh water that has just been brought to the boil to within 1 centimetre of the rim of the cup. Leave to stand for one minute. Stir, lightly mashing the teabag. Remove the teabag and discard. Add sweetener and stir again. Add a few drops of milk until the tea is opaque but not milky.

You might disagree with the details – there are many ways to produce a cup of tea (this happens to be how my wife likes it made) – but the words above represent a simple algorithm for making tea. An algorithm is a set of instructions to be followed to achieve an end.* Another familiar everyday form is a recipe: an algorithm for producing a particular food dish. Usually in the computing world algorithms are more about a set of logical operations that are to be undertaken on numbers or strings of text, but the concept is exactly the same.

* The word 'algorithm' comes from the Latin name Algorithmi, given to the ninth-century Persian mathematician al-Khwarizmi.

Sorting my books

Before we get on to the connection between algorithms and computer programs, and their importance for quantum computing, let's take another simple task and look at how different algorithms can improve the way that the task is completed. This particular task is a useful introduction as it's a real-world representation of a very common computing need.

I have a single Ikea 'Billy' bookcase that is full of non-fiction books. For reasons I can no longer remember, when I moved into my current house, I decided that it would be fun to put the non-fiction books on the shelves in any old order – randomly as I came across them. The rest of the bookcases are organised alphabetically by author,* but not the non-fiction one. However, after ten years of living with this randomness, I have become fed up with taking ages to find a specific book and have decided to reorganise them alphabetically by author's name. But how am I to go about that? What I need is an algorithm.

Let's start with what surely must be one of the worst approaches. I could take all the books off the shelves and randomly place them back. After all the books are back on the shelves, I check to see if they are in order. If I repeat that process often enough, I should eventually get them in the right order. The two problems with this are that it could take many, many attempts – and that I have to check every book to make sure that not a single one is out of order every time to know whether or not I have succeeded. I have just done a rough count and there are around 240 books in

* Except for the bookcase with books I've written and their translations, which are organised chronologically, as organising them alphabetically by author would be silly.

the bookcase. There are around 10^{58} ways – 1 followed by 58 zeros – to arrange 240 books, so this is likely to take me longer than the lifetime of the universe to complete. It's safe to say that this is not the algorithm to choose.

Let's try something more subtle: leaving the books on the shelves and looking just at the first pair. Are they in the right order alphabetically? If not, I swap them. (If they're both by the same author I check the date of publication and swap them if they are not in date order.) Then I look at books number two and three. Are their authors ordered alphabetically? If not, I swap them. And so on. I work through all the books, then go back to the start and do it all over again, with the exception that I don't need to check the final pair, because I've already shuffled the book whose author is the very last in the alphabet all the way to the end. Then I do the process again, but this time I can ignore the last two books. And so on. Eventually I will find every book is correctly placed with respect to its neighbour, and now I can stop.

It's a tedious approach, I admit – but computers are good at doing tedious things, and this is a genuine computer algorithm for sorting a list of items, known as a bubble sort. It's not a very efficient way of sorting data, but it's an easy technique to program, so has sometimes been used when speed isn't an issue. However, it's probably not the best approach for my books as in the worst possible case I might have to perform 239+238+237...+1 = 28,680 book swaps, which would take me a long time.*

* Working out the number of swaps here is surprisingly simple due to a nice trick often attributed to the mathematician Euler in his youth. A teacher is said to have given the class the task of adding 1+2+3...+100 as a way of keeping them busy. Euler came back with the correct answer in seconds. He had realised that there are

Okay, let's try another way. I take all the books off the shelves, then start to place them back one at a time. Whatever the first book is, I place it in the middle of the shelves. The second book I place to the left or right of the first one, depending whether or not the author comes alphabetically before or after the author of the original book. (As with all other examples, if both have the same author, I order them by date of publication.) With the next book, I again start with the middle book and work left or right as far as is necessary until my new author is to the left of the book with an author who comes after it alphabetically (or at the right-hand end). If I reach the end of a shelf I move up if going left and down if going right. And if a shelf is full, I randomly pick left or right – if left, I take the book from the left-hand end of the shelf and put it at the right-hand end of the shelf above. If right, I take the book from the right-hand end and put it at the left-hand end of the shelf below. If necessary, I repeat this process until there is space on a shelf. This is known as an insertion sort in computing, and in the worst case it's likely to take a similar number of actions as a bubble sort.

I can improve the speed of the algorithm by using a more sophisticated grouping. Here, I first take the books off the shelves in pairs and put each pair on a table, sorting all of the pairs into the right order. Then I take the first two pairs. I compare the first one of each pair and start a new group with whichever book comes first. I then compare the leftmost one remaining of each pair and add the first of these two to the new group. At the end of this process I have a group of four

---

50 pairs, starting with the first and last numbers, then the second and penultimate numbers and so on, all of which add up to 101 – so 1+100, 2+99, 3+98... etc. That makes the total 50×101 = 5,050. I did the same for 1+2+3...+238, then added on the unpaired 239.

in the right order. I do the same with the next four books, and so on. When all the pairs are chunked into fours (give or take the end one), I do the same with the groups of four, making groups of eight. This continues until all the books are in a single group, which will have been sorted into the correct order. Once again, this is a computer sorting algorithm, this one called a merge sort, devised by John von Neumann way back in 1945. This algorithm is significantly better than either of the previous examples: the worst case here should take 1,640 swaps to get into order.

This book sorting task is a genuine 'real world' problem – but has also given us a number of computer sorting algorithms, though there are many more available that are only really usable on a computer. Some of them make use of a particularly powerful aspect of a computing algorithm, known as recursion.

Hauled up by your own bootstraps

Recursion involves undertaking a repetitive task that acts on the outcome of the previous iteration of the task, plus a rule to start things off or stop them going on for ever. It's much easier to understand recursion by seeing it in action than by reading a description. Take, for example, the Fibonacci sequence, the famous sequence of numbers where the next number in the sequence is formed by adding together the previous two numbers. Here, the required starting rule is that the first two numbers are 1, as you need a 'previous two numbers' to make the sequence work. All it takes is repeated application of that rule to generate the sequence 1, 1, 2, 3, 5, 8, 13, 21 ...

The result is that a large outcome can be produced from a very small algorithm, a win for any computer programmer who has to produce the code. The kind of recursion used in generating the Fibonacci sequence goes on for ever – you stop whenever you like. But there is an alternative approach that can often be more useful of having a stopping value, which automatically causes the recursion to halt. So, for example, the largest odd factor* of a number can be worked out using the recursive algorithm: divide the number by 2. Repeatedly apply this to the result of the previous division until the result is odd.

For example, if my number is 624, applying the algorithm involves:

624 / 2 = 312; Is it odd? No.
312 / 2 = 156; Is it odd? No.
156 / 2 = 78; Is it odd? No.
78 / 2 = 39; Is it odd? Yes – so finish.

This may appear trivial, but in computing, recursion can be mind-twistingly powerful. To see why, we need to get a quick feel for what computer languages are and how they work.

Speaking to the computer

In a sense, the term 'computer language' is a misnomer, because computers don't speak these languages themselves. As we have seen, a computer handles 0s and 1s and nothing

---

* If your school maths is rusty, a factor of a number is a whole number that divides exactly into that first number.

else. In the early days of computing, these values had to be set directly, either by using switches on the front of a computer, or punched holes in either cards or paper tape. Here there was a direct physical analogue to the binary state in the computer. If there was no hole in the card or tape in a particular position it meant 0; if there was a hole it meant 1. Such binary instructions, provided in the form that the computer can use directly, are known as machine code.

This is not code in the sense used in codes and ciphers, where a code involves using a selected word or phrase to mean something else (so, 'SAUSAGE', for instance, could mean 'Advance to the border'). In general, the term 'code' is used in computing to mean the instructions as they are input into the computer, ranging all the way from machine code to 'computer code' or 'source code', which is usually something written in a higher-level, more human-friendly programming language (more on those in a moment).

Over time, an easier way of inputting instructions and data to computers was developed, known as assembly language. This was a set of very basic instructions to carry out an action, which was devised to fit with the way that the particular CPU handled data and operations, replacing the binary values for an instruction with an (often abbreviated) English word. So, for example, there might be an instruction such as MOV, followed by a location and value, to move a value into a particular location. Rather than writing them in binary, numbers in assembly language would typically be written in a more readable number system – though to keep easy correspondence with binary, the system was likely to require either octal (base 8) or, as is pretty much the standard now, hexadecimal. This means writing a number to base 16, so as well as the digits 0–9 there are additional values for the equivalents of decimal numbers 10–15, written as A, B, C, D, E and F.

Although a lot of early computer programs were written using assembly language, it was hard to write, easy to make mistakes in and particularly difficult to read afterwards to understand what was happening in the program when it needed to be changed or debugged. As a result, computer languages were developed that involved writing instructions in something even more easily understood by a human being, even if the instructions remained in a very strict format. This 'source code' would then be run through a program called a compiler, which turned it into assembly language, which would then be run through another program to generate executable code – the final result was a program that would run on the specific computer it was designed for.*

There were, and still are, a huge range of programming languages available, though variants of the C language, developed for the UNIX operating system in the 1970s, now dominate. Most rely on using a set of English keywords, such as IF or FOR, plus names for variables – effectively stores for holding numbers to be worked on – and a rigid syntax of symbols to indicate, for example, where a section begins and ends or whether a piece of text is part of the code or just a comment for humans to read, added to make the code easier to follow.

There have been some exceptions to the general trend. My favourite oddity was APL,** a language I learned in the 1980s, which was extremely compact and worked on whole

---

* Some very simple languages, such as the popular beginner's language BASIC, were able to run in an 'interpreted' mode, where they were effectively compiled as the program was run. This made them easier to use, but a lot slower to run.

** APL imaginatively stands for 'A Programming Language'.

matrices (multi-dimensional grids of numbers) at a time. It managed to keep things compact by using a number of extra characters that aren't in normal English usage, meaning it was preferable to have a special keyboard if you wanted to program in APL, but for data handling, and some specialist physics applications, it was phenomenally powerful. To give an idea of how compact APL was, the entire program to generate every prime number between 1 and N is:

```
(~N∊N∘.×N)/N←1↓ιN
```

As quantum computers are yet to have any of the standardisation we've seen in the mainstream computer industry, it's not entirely clear how they will be programmed as a matter of course. However, their special abilities, which we will go on to discuss, are neatly represented by matrices, and it may be that we see a resurgence of a variant of APL as a programming language for them.

The factorial dance

With an understanding of the basics of computer languages, we can now get a feel for why recursive code is so powerful. When we were looking at ways to sort my non-fiction books, I said that there were around 10^{58} ways to arrange 240 books. The exact number of ways to arrange the books, known as permutations in maths-speak, is 240! – pronounced '240 factorial'.*

---

* The exclamation mark symbol for factorial (entertainingly referred to as a 'shriek' in computer languages) was not universally welcomed by mathematicians when it was first introduced. Nineteenth-century English mathematician Augustus De Morgan is said to have moaned:

This is $240 \times 239 \times 238 \times 237 \ldots \times 3 \times 2 \times 1$. It is quite easy to see why 240! gives the right answer for the number of permutations if you think of the bookcase as having 240 slots, each of which can hold one book. For the first book I have a choice of any one of 240 slots. Once I have that first book in place in a specific slot, there are 239 other slots in which I can put the second book. And this is true for each of the 240 original slots – so there are 240×239 ways to organise those two books. And so on.

To write pseudo-code to produce a factorial recursively (it won't be in a real programming language, but my intention is to give the general look and feel of one), we need the concept of a function. Taken from mathematics, when we use a function in computing it is a black box that turns one thing into another. To use it, I don't need to know what's going on inside the box – though obviously someone does need to know this to write it in the first place. Often functions are written by someone else and provided in a library, so the programmer doesn't have to reinvent the wheel.

For example, I might have a function called 'Squared', which turns a number into its square. It would be written in my pseudo-code something like

```
number = 15
answer = Squared(number)
```

'Among the worst of barbarisms is that of introducing symbols which are quite new in mathematical, but perfectly understood in common, language ... the abbreviation n! ... gives their pages the appearance of expressing admiration that 2, 3, 4, etc. should be found in mathematical results.'

To use the function Squared, I first assign a value to a variable – I've decided to call that variable 'number', but it could be called 'x' or 'length' or whatever I like. The second line of my code then runs the function Squared, feeding the value of 'number' into it, indicated by putting it in brackets after the name. The Squared function finally assigns the appropriate value – in the case of my example 225 – to the variable called 'answer'. Inside the function, of course, would be the code required to multiply number by itself and return the outcome to the outside world, but when I use the function, I don't see that.

For our recursive example, we are going to produce a factorial function which I could similarly write as:

x=Factorial (n)

What this tells me is that my result (x) comes out of the black box called Factorial for the value n. For the example of my bookcase, n happens to be 240, but I could put in any number of which we want to find the factorial. Now let's dive inside the function. The code inside the function might look something like this:

```
Factorial (n)
If n equals 0, return 1
Otherwise return n × Factorial (n–1)
End
```

Let's see what happens if we set n to 240. Here n clearly doesn't equal 0, which means the function wants to return (i.e. send out to the rest of my program) 240 × Factorial (239) – whatever that is. So, the Factorial function sets itself going a second time on the smaller number. All this happens over and over again, with more and more instances of the Factorial function, until n gets down to 0. Then the results ripple back up to the original Factorial (240) – which comes to an end, giving us the final huge number.* Perhaps it's because I have a background in programming, but I find it truly beautiful that such a tiny piece of code, rippling values back up to the top, rather like the repeated image that you get when you stand between a pair of parallel mirrors, can produce such an impressive result.

Going in deep

So far, we have got a feel for the basics of how computer hardware and software do their jobs in the kind of conventional computer that you might have on your desktop, or in your pocket as a phone. When these devices operate, they are entirely dependent on quantum physics because electronics (as the name suggests) is all about the behaviour of electrons. These tiny particles, which are found around the outside of atoms and which flow through wires as an electrical current, are quantum particles. This means they obey the laws of quantum physics – they behave in ways that are surprisingly different from the familiar objects around us, even though those familiar objects are themselves constructed from atoms, which are also quantum particles.

* In practice it would probably have run out of space to do its calculation – the result is a 59-digit number after all – so a real piece of code to perform this function would have to check for things going wrong. This is why, if you ask Google what 240! is, it says 'undefined' rather than crashing horribly (fun though that might be).

However, conventional computers do not make use of quantum strangeness, except in special physical devices such as flash memory, which use quantum effects to keep an electrical charge stored when the computer's power is switched off. Before we can see how quantum computers are able to do apparently impossible calculations, we need to get a feel for what that quantum strangeness involves.

QUANTUM STRANGENESS

4

Immensely simplified, quantum physics has two rules:

1. Very small things don't have locations, we just have probabilities of where they are.
2. The first rule only works if these very small things don't interact with their environment.

As we shall see, once we include a bit of detail, the reality is distinctly more subtle than this – and there are other aspects at play – but these two features of the world of the very small make it behave entirely different from the world we observe on the scale of objects we can see with the naked eye.

This is conventionally interpreted as the quantum world being strange – and it is strange, from our distinctly parochial viewpoint. But this strangeness appears to be how reality is, so we need to get used to it. As the great twentieth-century quantum physicist Richard Feynman once said, 'I hope you can accept Nature as she is – absurd.'

Cosmic zoom

There was something of a vogue in the 1950s and 60s for the idea of zooming in from an overview of galaxies, stars and planets on down to a smaller and smaller world, until atoms themselves become visible, like solar systems in their own right. In the (surprisingly intelligent) 1957 film *The Incredible Shrinking Man*, the main character Scott, played by Grant Williams, starts shrinking after he passes through a mysterious fog. At the end of the film, after the inevitable battle with tiny predators, Scott has become philosophical about his fate, telling us that the infinitesimal and the infinite are 'the two ends of the same concept'. The film's viewpoint reflected, it seems, the beguiling idea of the atomic structure, with electrons whirling around a central nucleus, providing a direct and meaningful parallel with the structure of the solar system or of a galaxy.

A few years later in 1968, Eva Szasz made a famous eight-minute film called *Cosmic Zoom*, which starts with a picture of a rowing boat on a river, zooms out to take in the solar system, the Milky Way and the universe at large, then zooms back in to reach the structure of an atom before returning to the original view. Again, there is the feeling of continuity, that it's all part of one great, structured whole, where the largest and the smallest aspects are meaningful echoes of each other.

While there is no doubting that atoms are part of the universe, for over 100 years now, we have been well aware that things work very differently on the scale of atoms than they do for the everyday objects around us, or for planets, solar systems and galaxies. Thanks to the work of Max Planck, Albert Einstein and Niels Bohr, theory that would form the initial foundation of quantum physics, by 1913, the picture of the atom as being like a miniature solar system had been totally shattered.

It was only at the end of the nineteenth century that atoms started to be accepted as real things, and widely accepted proof really only came as late as 1905. So, for the majority of the time that atoms have been accepted as existing, we have known that they are very different from the kind of graphical representation of little blobs orbiting a bigger blob that is still commonly used today.

In reality, there is very little similarity between the behaviour of objects on the scale of atoms and the world around us that we directly observe. Pretty well everything we know, from light to whatever object you are using to read this book, is not a continuous thing but rather is made up of tiny components, which for want of a better word we tend to call particles. We're stuck with the term, but those particles are nothing like the specks of dust or powder that the name suggests – they have their own, special quantum reality.

Things come in chunks

The word 'quantum' (plural quanta), meaning an amount or quantity of an entity, is used scientifically to refer to the smallest chunks into which we can divide something.* Often

---

* Sort of. Some of these chunks can themselves be made of smaller chunks – so, for example, atoms are quantum particles despite having a structure of smaller particles – but it's a good approximation.

those chunks are referred to as particles. A particle* is a small part of something, but we tend to think of it as referring to a tiny grain or piece of stuff. The trouble with using this term for quantum particles – particles so small that they obey the rules of quantum physics – is that these things behave nothing like a grain of salt or a dust mote.

Normal, familiar particles might be tiny, and so able to float around in the air if light enough, because the air molecules keep bashing into them, but such a particle behaves just like a smaller version of a ball or a stone. However, quantum particles obey totally different rules. Which is confusing, because so many everyday and familiar things are made up of these strangely behaving particles.

As we have seen, atoms and molecules were pretty much proved to exist in 1905, which was thanks to a clever piece of work by Albert Einstein that mathematically modelled the interaction of molecules with small particles in liquid suspension, known as Brownian motion. The same year, Einstein also showed that light came in chunks. This was significantly more of a shock. Although a piece of matter looks continuous, it seemed quite reasonable that it should be made up of smaller components. Even some of the Ancient Greeks had suggested the existence of atoms, the limit of cutting something up as small as possible. But by the start of the nineteenth century, light had been proved to act like a wave.

---

* The earlier term, used, for example, by Newton was corpuscle, meaning a little body. It was still more common at the end of the nineteenth century when the physicist J.J. Thomson showed the existence of what we would now call electrons and referred to them as corpuscles.

A wave is a continuous, moving disturbance in something – think, for instance, of a wave in water or sent down a piece of string. Light has clear behaviour suggesting that it is a wave, though it proved stubbornly difficult to discover what the 'something' was in which light caused a disturbance. Nevertheless, light did things, such as refraction (changing direction as it passed between media in which it travelled at different speeds), that seemed to prove it was a wave.

At the turn of the century, the German physicist Max Planck had come up with the idea that it would be convenient if light came in chunks to explain how hot things glowed with the colours they do. Planck didn't believe this was true – everyone knew light was a wave – but the theory of the time predicted that even at room temperature, objects should glow brightly in the blue and ultraviolet. However, assuming that light was not a continuous wave, but came in what Planck called 'quanta', made the maths work and match observation. Soon after, Albert Einstein went one step further than Planck. Einstein came up with an explanation of a strange phenomenon known as the photoelectric effect, which would only work if light really did come in chunks – chunks that would later be known as photons.

In the photoelectric effect, an electrical current is generated in some metals simply by shining a light onto it. Observations indicated that the light was managing to knock electrons out of the metal, starting the current flowing. However, the effect relied not on the brightness of the light, but its colour, which reflects the amount of energy in individual photons of the light. Make the light too red (too low in energy) and it didn't matter how bright it was, it wouldn't knock out any electrons. This didn't make sense for the continuous flow of a wave, as bigger waves carry more energy. But it would happen if light came in chunks (photons), and a single photon was responsible for knocking a single electron out of the metal – that photon would have to have enough energy to do the job, meaning it would need to be nearer the blue end of the spectrum or above.

When the young Danish physicist Niels Bohr came up with a first quantum theory for the structure of the atom, he made use of Einstein's idea. Atoms absorb and give off light – and they tend to do so with specific colours (energies) depending on the atom in question. This is how the science of spectroscopy works, which enables us to tell which elements are in a material by the colours of light it gives off when it is heated, or the colours it absorbs when white light passes through it. This is the reason, for example, that sodium lamps have a distinctive yellow-orange hue. Bohr realised that these bands of colour could reflect the electrons in the atom being restricted to specific-sized changes in energy, so-called quantum leaps. The electrons receive or lose energy when light hits them or they give light off, but can only do so with particular, quantised changes in energy.

This made double sense, as electrons couldn't be allowed to orbit around the nucleus of an atom like a planet around the Sun, because electrically charged particles give off energy when they are accelerated, and orbiting involves constant acceleration. The only way Bohr could make an atom stable was by keeping the electron at a particular distance from the nucleus as if it were on tracks. It could jump from one track to another (hence the quantum leaps), but it could not occupy any position in between the tracks.

This was the beginning of the quantum revolution. But it would take the work of two further quantum physicists to introduce the most mind-boggling aspects of the behaviour of very small things.

That equation (no cats involved)

One essential step forward in quantum theory emerged in the form of Schrödinger's equation (there are various alternative formulations and extensions, but we'll just concern ourselves with the basics). The work of Austrian physicist Erwin Schrödinger, the equation, or to be precise its square, gives, for example, the probability of finding a quantum particle in a certain location.

At first sight, what it says is bizarre. If we think, for example, of a familiar object like a ball being thrown from a hand to hit a wall, we can predict (given some details and basic maths) where the ball will be at every point along its trajectory.* If we imagine undertaking this exercise with a quantum particle, Schrödinger's equation tells us the probability of finding the particle at any location – and, to be fair to Newton, it is relatively likely that the particle will be discovered on the kind of path you would expect it to take. But there are also probabilities of finding it all over the place, well away from that path.

According to Schrödinger, until the particle interacts with its environment, for example causing a detector to ping, we can't say where it is. To be honest, when Schrödinger came up with his equation there was relatively little experimental

* This is a physicist's ball – a perfect sphere with no air resistance, etc. to throw it off its trajectory as expected by Isaac Newton. It gets quite difficult to be precise with a real ball, but the physicist's ball is close enough to reality for our purposes.

data to back up this outrageous claim. (To be precise, the claim was partly down to Max Born, a friend of Einstein's, who was the one who suggested the outcome of the equation was a collection of probabilities, rather than describing the actual location of the particle.) However, since the 1920s, Schrödinger's equation has been totally vindicated. It's how things are, whether we like it or not.

This, incidentally, has nothing to do with Schrödinger's rather more famous cat – which we will briefly come back to under the topic of superposition (see below). Personally, I find this over-exposed moggy distinctly irritating. However, the equation has everything to do with our existence and some of the impressive abilities of our phones. This is because Schrödinger's equation tells us to expect something called quantum tunnelling.

Usually, if we want to get to the other side of a barrier without going over it, under it or around it, we have to go through it.* This is what we usually mean by tunnelling. However, the probabilistic locations given by Schrödinger's equation give us another, stranger option: not actually tunnelling through the barrier, but already being the other side of it without ever passing through it. This is something quantum particles are entirely capable of doing. It's as if you park your car in the garage one night and next morning it's sitting in the drive, without the door ever being opened.

Quantum tunnelling is relatively unlikely to happen. The probability is typically very low. But because there are often a whole lot of quantum particles in any particular object, it

---

* Any parent will probably realise that this could be called going on a quantum bear hunt.

happens sufficiently often to have impressive effects. One place it happens is in the Sun. For the Sun to work and provide the heat and light that keep us alive, positively charged quantum particles called ions inside it have to get very close together in order to make it possible for the nuclear fusion process that generates the Sun's energy to take place. Because the ions all have the same charge, they repel each other – and as they get close together, that repulsive force becomes very strong. So strong that even in the heat and pressure of the Sun, the ions don't get close enough to fuse. But because a small percentage of the ions tunnel through the barrier of the repulsion, the Sun functions and we stay alive.

On a distinctly smaller scale, the flash memory that stores information in your phone or on a memory stick even if there is no electrical power applied to it makes use of quantum tunnelling. As we have seen (page 22), computer memory holds information as electrical charges, as do solid-state drives (SSDs). However, when the power is turned off, the charge in memory dissipates.* But flash memory holds the charge representing data in an insulated space. This stops the values in the bits being lost – but the price is that these bits are difficult to get to, to change the values or to read them. It's by using quantum tunnelling to enable electrons to pass through the barrier and interact with the stored bits that the computer or phone's circuits can access the memory.

* The charge doesn't necessarily disappear instantly, which is why the light often doesn't go off for a second or two on some chargers and other electrical devices when you unplug them. But it doesn't hang around for long.

Being uncertain

The other famous name from the 1920s period in the quantum world was the German scientist Werner Heisenberg.* His name is firmly associated with the uncertainty principle. This is probably the most misused concept in all of quantum physics, a discipline that is routinely misused by taking its terminology and pretending the words justify all sorts of totally fictitious concepts that have nothing to do with physics.**

The uncertainty principle does not say that 'everything is uncertain' or that 'anything goes'. In fact, it is a clear mathematical statement. It reflects the way that different aspects of the physical world are intimately connected at the quantum level. Its best-known formulation is that the more accurately you know the position of a quantum particle, the less accurately you know its momentum.*** It is impossible to know both perfectly at the same time.

The uncertainty principle also applies to a number of other pairs of properties of a quantum particle or quantum system, most notably the combination of its energy and the time interval over which it is observed. If you take a very small span of time, there is a very wide uncertainty in exactly

---

* For completeness we ought to include others such as the British physicist Paul Dirac, and the French Louis de Broglie, but Heisenberg is as far as we need to go for much of the quantum strangeness that will be relevant to quantum computers.

** This can be anything from Quantum dishwasher tablets to suggestions that the human body is just a holographic projection of the mind, so simply thinking the right thoughts – indulging in quantum thinking – can produce quantum healing. In all these cases, the word 'quantum' could just as easily be replaced by 'magic'.

*** Momentum is mass times velocity – effectively it represents the 'oomph' of movement that a particle has.

how much energy the system has. As a quantum system can include apparently empty space, this means that the energy level of empty space varies considerably over extremely small intervals of time, so much so that the energy can briefly be enough to take the form of particles (mass and energy being interchangeable, as indicated by Einstein's famous $E=mc^2$ equation). These 'virtual particles' cannot be directly detected as they disappear so quickly, but their existence can be indirectly detected.

Superposition

We're beginning to get a picture of quantum particles as being very strange indeed. And the lack of a concrete position that we discovered in the trajectory of a quantum particle is reflected in the probabilistic nature of other properties of such particles. 'Properties' is a term widely used in the quantum world, so it's probably worth making it clear what's meant. A quantum particle's properties are things like its position, its mass, its electrical charge, its spin (more on this in the next section) and so on. Some of the properties, such as charge, do not vary. But the interesting ones, for this purpose, are the variable ones, and these are the ones that, when not measured, typically only exist as probabilities rather than specific, fixed values.

Central to understanding this is realising that these probabilities are applied in a very different way to the way probabilities tend to be used in everyday life. Take what is surely the most familiar application of probability – the toss of a (fair) coin. We say that there's a 50 per cent probability of getting a head and a 50 per cent probability of getting a tail. What we mean by this is that, over a long enough sequence of tosses, 50 per cent of the results should be heads and 50 per cent tails. But let's take a specific coin. I toss it and leave my hand over the top of it. What is the chance that the coin is lying heads up?

We would usually say that it is 50 per cent, because the chance of tossing a head is 50 per cent. Strictly, though, we shouldn't say this about the coin. That coin is either showing a head or a tail with 100 per cent certainty. We just don't know which is the case until we look at it. It would be more accurate to say that on repeating the experiment many times, we would be right 50 per cent of the time if we said the top face was a head. The actual head or tail result under my hand is what's known to quantum physicists as a hidden variable. It is a real value, but we just don't happen to know what it is.

Einstein, and those who followed his approach, thought that quantum particles had hidden values too. So, although Schrödinger's equation tells us the probability of finding a particle in various locations, Einstein believed that the particle had an actual, exact location – it just wasn't known until it was measured. Einstein's opponents, notably Niels Bohr, argued that, on the contrary, all that existed were the probabilities until the particle interacted with its surroundings – for example being 'observed' by a detector. There were no hidden values. It was only on being observed that the probabilities became an actual value.

This is sometimes described as a particle being in two places at the same time – but that's a very timid way of looking at things.* The reality is more that the particle has

---

* The 'two places at the same time' description almost certainly comes from a very widely used experiment called Young's slits, where a quantum particle has the opportunity to go through one

simultaneous probabilities of being in all its different possible positions – and all that exists is those probabilities. This ability to have multiple simultaneous probabilistic values is known as superposition. When we start to look at how quantum computers work, this effect will be concerned with properties such as quantum spin and polarisation, where superposition has a crucial role to play in the computers' remarkable capabilities.

You spin right round (and also don't)

Quantum spin is perhaps the most confusingly named of the properties of a quantum particle. It is called 'spin' because of some similarities to angular momentum – the degree of rotational 'oomph' a spinning object has. However, in practice, quantum spin does not represent spinning around in the conventional sense at all. When the quantum spin of a particle is measured it can only be either 'up' or 'down' in the direction it is measured. Remember that 'quantum' means coming in chunks – here the chunking has only two possible values, up or down.

The quantum nature of spin was first discovered in a classic investigation from the 1920s known as the Stern–Gerlach experiment, named after Otto Stern who came up with the idea and Walther Gerlach who undertook the experiment. A stream of silver atoms was passed through a strong magnetic field, which was not symmetrical due to the shaping of the magnets. As silver has a single electron on the outside, the

---

of two separate slits, but having a probability to go through each of them is said to be in two places at once.

expectation was that each atom would act as a little magnet, and would be deflected differing amounts by the magnetic field, depending on the direction in which its magnetic poles were aligned. What was observed, though, instead of seeing a continuous range of deflections, was that all the atoms were either deflected up or down by the same amount.

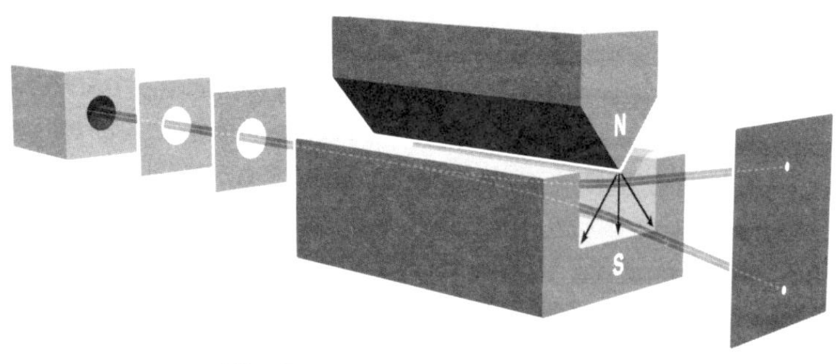

The Stern–Gerlach experiment.

This experiment emphasises how much quantum spin is *not* about spinning around – it is a magnetic effect, but quite unlike the behaviour that would be expected from conventional magnets. The Stern–Gerlach experiment makes a measurement of 'spin' in the up/down direction as defined by the orientation of the magnets. Although the original experiment involved silver atoms, the same effect is observed with electrons and with charged ions,* which are likely to be more of interest in quantum computers.**

---

* Ions are atoms that have gained or lost one or more electrons and so have an electrical charge.

** The simple Stern–Gerlach experiment can't actually be done on electrons, as unlike a silver atom, they're electrically charged, and moving electrical charges are also affected by magnets; but by using

The outcome of measurement, though, does not reflect the state of the particle before that interaction with its environment. Before a measurement is made, the particle is said to be in a superposition of its states – both up and down at the same time. It doesn't have to be equally likely that the outcome on measurement will be up or down. It might have, say, a 70 per cent probability of coming out up and a 30 per cent probability of coming out down when measured in a particular direction. But just as only probabilities exist for a particle's location before measurement, so in the superposed state, only probabilities exist for a particle's spin.

It was the absurdity of the binary superposition that is shown, for example, in the up/down probabilities of spin that Schrödinger was pointing out when he came up with his infamous cat 'experiment'. The idea of the thought experiment was to use a different binary quantum outcome, in this case whether a radioactive atom had decayed or not. The chances of this happening after a particular length of time are probabilistic. So, after that time has elapsed, if the atom hasn't interacted with its environment, the particle exists in a state of probabilities for the two options of decayed or not decayed.

Where the cat example gets silly (it is described as an experiment, but it could never be carried out in reality), is that Schrödinger had a detector 'watching' the particle. When the particle decayed, the detector would register this and release a poison gas into a box containing a cat. So, after a while, before the box was opened, the cat was said to be both alive and dead. The reason it's silly is that the particle's

---

more sophisticated equipment, the same effect can be detected using electrons.

interaction with the detector stops it being in a superposed state. The whole idea was not that the experiment could ever be carried out, but that it made you think about the absurdity of superposition. However, as Richard Feynman said, nature is absurd. It's what happens.

Polarisation

A second quantum property that is commonly used in experimental quantum computers is polarisation. Like spin, this is a property that is associated with a particular direction. In conventional polarisation, a photon is polarised in a particular direction, or alternatively the direction of polarisation can change progressively with time in so-called circular polarisation. When taking the traditional wave approach to light, polarisation is the direction the electrical wave oscillates, but here we are taking a quantum approach, where polarisation is the equivalent of spin for a photon. Some materials, most famously the commercial product Polaroid, act as filters for polarised light, only letting it through if it is polarised in a particular direction. This is why Polaroid sunglasses reduce glare – because reflected light is typically more polarised in one direction, where ordinary light has photons polarised in every direction. The lenses of the sunglasses block light polarised in the direction typical in reflections from horizontal surfaces.

Just like spin, the polarisation directions of photons tend to be superposed. This can be shown with experiments using polarising materials.* If we first pass photons

* You can try this at home with the lenses from polarising glasses, such as those used to watch 3D films.

through a polariser at 45 degrees between vertical and horizontal, then through a vertical polariser, and polarisation were a conventional property, we would expect no photons to emerge the other side – none of them would have been vertically polarised. But in reality, after passing through the 45-degree polariser, the photons are put in a 50:50 superposed state of vertical and horizontal polarisation. This means that half of the photons will make it through the vertical polariser.

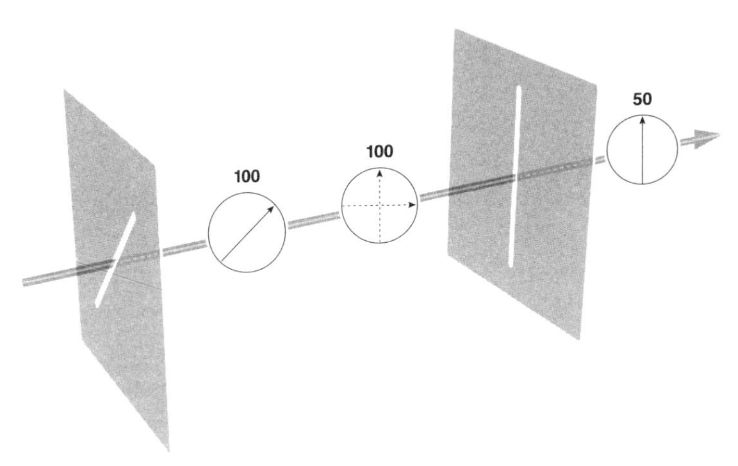

If we send 100 photons polarised at 45 degrees, 50 of them (on average) will get through the vertical polariser.

It might seem that this could be a result of half of the photons being vertically polarised and half being horizontally polarised, rather than each photon being in a superposed state, but by using three polarisers we can prove that this isn't the case. If we start with two polarisers at 90 degrees to each other, nothing gets through.

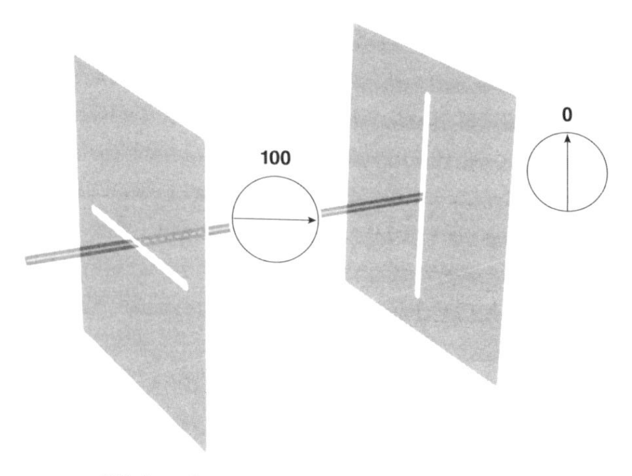

With polarisers at 90 degrees to each other, nothing gets through.

But if we insert a polariser in between the other two at 45 degrees, some photons do get through to the other side.

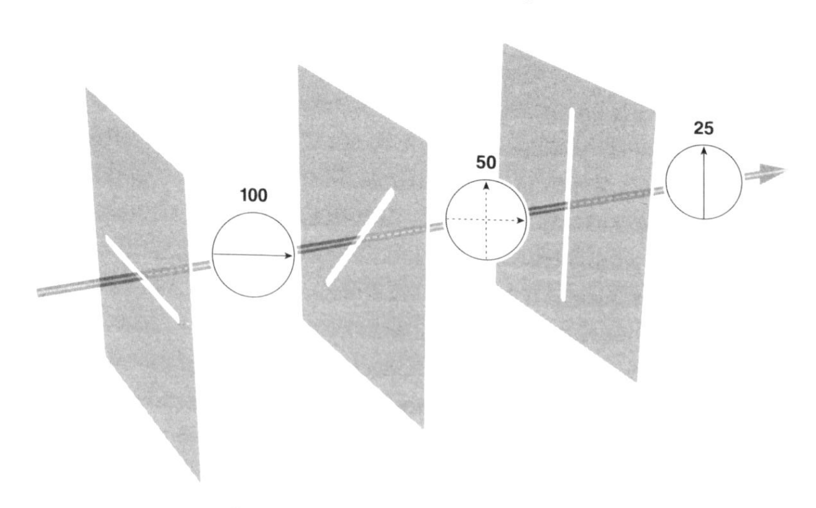

An angled polariser in the middle allows some photons through.

If the 45-degree-angle polariser let through a 50:50 mix of horizontal and vertically polarised photons, then we would expect that the photons that got through it would all still be horizontally polarised and would be stopped by the vertical polariser. But because the angled polariser produces photons in a superposed state of horizontal and vertical, some photons will get through to the other side.*

Entanglement

There is one last significant aspect of quantum physics called quantum entanglement, which was brought to the fore in a paper by Einstein and two colleagues. It will prove essential to understand this to get our heads around quantum computers, as it is vital to connect parts of a quantum computing device.

The idea of entanglement had been around a few years when Einstein's paper came out. If quantum theory were correct, the ability to entangle particles was predicted. This meant that it should be possible to produce quantum particles in a way that they have a special link, so much so that as long as the entanglement remains, a change to one particle is instantly reflected in the other, however far apart the pair is separated. The simplest way to produce such particles is if a pair of photons of light are produced by the same electron

* Incidentally, the three-polariser experiment is reminiscent of the way that a liquid crystal display works. The display has a backlight, then a horizontal polariser (say), then the liquid crystal, then a vertical polariser. Left to its own devices, the liquid crystal has no effect on the polarisation, and no light gets through. But put an electrical voltage across it and the liquid crystal rotates the polarisation, so light starts to get through to the front of the screen.

dropping to a lower energy level in an atom, but there are also a number of techniques available to entangle existing particles.

Einstein's paper from 1935, known as EPR after its authors (Einstein, Podolsky and Rosen), though formally titled 'Can Quantum-Mechanical Description of Physical Reality Be Considered Complete?', provided a thought experiment to explore whether or not quantum physics was correct. It envisaged producing two entangled quantum particles which fly off in opposite directions. After a while, simultaneous measurements are made on a pair of properties of the particles. According to quantum theory, until a measurement was made, these properties would not have set values, just probabilities. But as soon as the measurement was made, a specific value would be produced. And what was discovered for one particle would *instantly* imply a specific value for the other. But how could this be the case, as nothing, not even information, was supposed to be able to travel faster than light?

So, for example, one of the properties discussed in the paper was the momentum of the particles. The exact value was unknown – it could have a whole range of values with varying probabilities. But as soon as the momentum of one particle was measured, the other particle had to have exactly the same momentum in the opposite direction, as a law of physics, the conservation of momentum, required this.

In practice, the EPR paper caused considerable confusion, because it mentioned two different properties, momentum and position. As we have already seen, these are paired properties in the uncertainty principle, and a lot of people thought that in some way Einstein was proposing an experiment where both the momentum and position could be measured perfectly simultaneously, breaching the uncertainty principle. In reality this was not the intent. When asked about measuring the two properties, Einstein commented, *'Ist mir Würst'*, literally 'It's sausage to me', meaning 'I couldn't care less'.

The point of the thought experiment was not about the simultaneous measurement, which in practice could not be performed, but rather that either of the measurements would somehow instantly influence the other particle at any distance. Later versions of the thought experiment used the single quantum property of spin we've already met (see page 59). If one particle was spin up when it was measured, then the other particle would instantly become spin down.

Einstein's argument was that either this meant that the particles already knew how they would end up before they were separated (so-called hidden variables, like the coin that has already been tossed) – we just didn't know what the answer was – or that the two particles could influence each other remotely, which went against the idea known as locality. This said that you could only influence something which you had contact with, whether directly (as when I push you) or indirectly (as when I say something which pushes waves through the air, which pushes against your eardrum). As moving away from locality seemed ridiculous to him, Einstein deduced that quantum physics was incomplete – that there were hidden variables we didn't know about.

Making entanglement real

For many years, the EPR experiment was just a hypothetical example. There was no way of testing it and interest in it faded. By contrast, quantum physics stubbornly proved extremely good at describing reality and was increasingly of practical use. All the electronic devices that started to be possible within a couple of decades of Einstein's paper could not have worked if quantum theory did not give an accurate way of predicting how real electrons and photons behaved. In a sense, it didn't matter if the reasoning behind it was 'right' – the theory produced the right numbers. It worked.

However, in the 1960s, the Northern Irish physicist John Bell developed a practical measure that could be used to see which of Einstein's possible outcomes was correct. A decade or so later, French physicist Alain Aspect devised the first of a series of experiments that have repeatedly proved that it is locality that breaks down rather than quantum theory being wrong – it really is possible for one of a pair of entangled particles to instantly influence the other at any distance.

These experiments had to be quite clever: there had to be a way to make sure that there wasn't time for information to get from one particle to the other at the speed of light – which in the short distances inside a lab and considering the extremely high speed of light meant being able to make extremely rapid changes to the experimental setup. But time after time the experiments backed up entanglement – which Einstein once called 'spooky action at a distance' – as being real.

The latest versions of these experiments go far beyond the limits of the laboratory. For example, in 2017 a Chinese satellite called Micius, named after an ancient philosopher, sent entangled pairs of photons down to Earth locations 1,200 kilometres apart, making it possible to demonstrate the effect over a considerable range where there could be no possibility of interaction between the two photons in the pair. It's not easy to do this – quantum entanglement is fragile, as should a particle interact with its environment the entanglement will be lost – but with enough particles it is possible to maintain the entanglement for a small percentage of those sent out.

Entangled time machines

It might seem that the most exciting way to use entanglement would be to send a message instantly from place to place. Were it possible, this would certainly be extremely useful. Without entanglement, our communications are limited to the speed of light. At 299,792,458 metres per second (in a vacuum), that is pretty fast. From location to location on the Earth, light speed feels close to instantaneous. However, for computers, the kind of sub-second delays this introduces over even short distances can cause problems, and if we are sending messages to satellites or, for example, to a base on Mars, even light speed can result in considerable delays. The time to send a message from Earth to Mars varies depending on the relative orbital positions of the two planets, but it can take well over 20 minutes each way.

More interesting still – mind-bogglingly so – if you could send a message instantly, it would also be possible to send a message backwards in time. This gets a bit messy in practice, but the basic setup is relatively simple and relies on Einstein's special theory of relativity. This much-confirmed theory says that when an object is moving, that motion has an effect on the passage of time (and various other things). For example, if a spaceship is travelling away from the Earth at high speed, from the point of view of the Earth, the time on the ship is running slowly. This means that after travelling for a while, significantly less time has passed on the ship than has on Earth.

This is not just an apparent but unreal effect, like the way an object looks smaller if it's further away from us. It is a genuine physical phenomenon. Time really does go slower on the moving ship as seen from Earth. Experiments have been done transporting an atomic clock around the Earth – when the clock got back to base, less time had elapsed for the travelling clock than for an identical one that had stayed put.*

Now, imagine we've got a ship that has been on its way for some time, flying away from Earth as fast as possible. From the Earth, the time on that ship has run slowly. The ship has moved into the past. If we are able to send an instantaneous message to it using entanglement, it will arrive before we sent it. But here's where things get interesting. This situation is symmetrical. From the viewpoint of the travelling ship, the time on Earth has been running slowly. From the ship, the Earth is in *its* past. So, if the message is then relayed instantaneous back to Earth, it will arrive where it started from before it was sent. The message has travelled into the past, not on some remote ship, but on arrival back home.

While at the moment we couldn't achieve much in the way of time differentials because our rockets don't travel very quickly compared to the speed of light (our best time machine yet using this effect is Voyager 1; moving away from Earth since 1977, it has travelled around 1.1 second into the past), it still would be an incredible feat if instantaneous

* The first time this experiment was done they couldn't afford their own transport, so had to strap the bomb-like clock into a seat on a commercial passenger plane. It seems unlikely this would happen today.

communication were possible. But all the evidence is that it's not going to happen. This is because the information conveyed by entanglement is the outcome of a random process. We don't know what the result of taking a measurement will be until we do it – and we can't control the outcome. For decades, physicists have played around trying to discover ideas of mechanisms to use entanglement to send a message and they have all failed.

The entangled crown jewels

The inability to send a message into the past is disappointing* – but experiments like the Micius satellite would not be worth the investment if entanglement didn't have some practical applications. And it does.

The easiest one to make use of, and one that satellites like Micius are likely to provide the infrastructure for, is quantum encryption. The encryption we currently use on computers – the one used, for example, when you use SSL security on a website (represented by a padlock in the address bar of the browser) – is not unbreakable. It's just very, very hard to break. But, as we will see in more detail later (page 81), quantum computers have the potential to be able to be able to break such encryption at some point in the future. However, there is a totally unbreakable form of encryption, and it has been used for around 100 years.

The unbreakable method is called a one-time pad. Encryption often involves a key – a set of characters whose values are added to those of the message to produce the

* Unless you run a lottery.

enciphered result. So, for example, if my message is 'HELLO' and I use as key 12345, I move 1 step along the alphabet from H, making it I, I move 2 steps along from E, making it G and so on, producing the enciphered text IGOPT. To decipher the text, I subtract the key and get back to HELLO.

In a one-time pad, the characters of the key are not a simple sequence like my 12345 but are randomly generated, which means that the encrypted text is unbreakable. There is no pattern in the key which will enable a codebreaker to deduce what it was. However, the reason this system isn't as widely used as breakable methods is that it has a serious flaw. To be able to use it, you have to get the key safely to both the sender and the receiver of the message. And that means that the key is vulnerable to interception.

However, in the properties of entangled particles we have a naturally random set of values, which aren't generated until the key has already been sent to both sender and receiver – so as long as the particles stay entangled until they are used, the encryption is unbreakable.* The ability to provide quantum encryption will become increasingly valuable as quantum computers are introduced.

Before that can happen, though, quantum computers have to be built – and that can't be done without making use of another application of entanglement, known as quantum teleportation. Acting a little like a miniature version of a *Star Trek* transporter, quantum teleportation involves transferring one or more properties from one quantum particle to another. Under normal circumstances this isn't possible,

* To make quantum encryption work there have to be extra checks, for instance to ensure that entanglement hasn't been broken before the values are used, but if used properly quantum encryption has all the benefits of the one-time pad and none of the drawbacks.

as looking at a quantum particle to discover its properties changes it from being in a superposed state to having a single 'actual' value. But to get information into and around a quantum computer, as we shall see, we need to be able to make use of a particle's superposed state without observing it.

The clever thing about quantum teleportation is that no one ever discovers the value of the property. The superposed state is transferred from one particle to another, and this is possible because it is never 'looked at'. As a result of the process, the original particle loses its state – in the process of making what amounts to a remote copy, the original is scrambled. But this does provide an essential mechanism for handling the trickily sensitive quantum particles that will be used as the equivalent of bits in a quantum computer.

We have our building blocks

With an understanding of the basic workings of computers and their logic gates, plus a feel for the strange behaviour of quantum particles, we are ready to bring the two together to see how a quantum computer works and why it has the potential to be so much better than a conventional computer for certain (but definitely not all) applications.

More often than not when thinking about computers we would consider the hardware first – how the physical computer works – and then add in the software. That's exactly what we did when finding out more about conventional computers. But as is often the case, quantum computers turn things on their heads. As we have seen, quantum algorithms already exist, well before quantum computers were able to run them. So, let's dive into a couple of those algorithms in some more detail.

QUANTUM ALGORITHMS 5

In the opening chapter (see page 5) we met Lov Grover and his remarkable quantum search algorithm. As we discovered, this would enable a search of unstructured data that required a program only to examine the square root of the number of entries in a database, unlike a traditional search which could in principle require all the entries to be checked, and on average would need to check half of them.

The mathematics and physics of the algorithm is a little fiddly to get your head around. (If you are reasonably familiar with quantum physics, take a look at the paper, referenced in the Further Reading section.) However, the principle that Grover describes makes use of the strange nature of quantum interactions. He points out that quantum mechanical systems can make what he calls 'interaction free measurements'. The idea is that a quantum particle can be in a superposed state where it has a probability of interacting with more than one object.

If we can set up such a superposition, it makes it possible to detect the thing you are looking for without looking in each location, but rather by using a quantum particle which has simultaneous probabilities of looking in a range of locations. So, Grover's algorithm involves three steps. First the system is set up in a superposition, then a number of operations are undertaken that involve changing the state of the system as many times as the square root of the number of search locations. The specific state change depends on a condition determined by a quantum gate operation (more on quantum gates in the next chapter). After this, a measurement is made on the final state. Being a quantum computer, we can't get a definitive outcome, but in over 50 per cent of cases it will point to the correct item – if not, it's necessary to repeat the process. But overall it is still far quicker than searching through every item in the database.

It's pretty obvious how this kind of ability to search using the square root of the number of operations has the potential to speed up the function of a search engine or a database – which as we have seen is bread-and-butter software in computing, used for everything from accounts to airline ticketing systems. However, a few years later, Grover came up with a second quantum search algorithm that could extend searching from its basic mechanical nature to something far more like the way the human brain searches through memories. To see how this works we need a quick reminder of Boolean logic.

Professor Boole's logic

When we explored the working of gates in conventional computers, we found that their actions could be represented by terms such as AND, OR and NOT. These basic operations did not originate with computers but date back to the work of a nineteenth-century, mostly self-taught, English mathematician by the name of George Boole. Without doubt, Boole's greatest contribution was in the field of symbolic logic.

The systematic application of logic goes back to the Ancient Greeks, but for them logic was all about the truth or falsity of statements. An example of this would be deduction. So, for example, we might have the following statements:

All pairs of scissors have two blades.
This object has one blade.
This object is not a pair of scissors.

The final sentence 'This object is not a pair of scissors' is logically *deduced* from the previous ones. Such verbal manipulation can be entertaining, though it can also be a little dangerous. For example, if we tried:

All pairs of scissors have two blades.
This object has two blades.
This object is a pair of scissors.

we would have made a logical error in our deduction. While it's true (according to my opening statement) that all pairs of scissors have two blades, this doesn't automatically reverse to mean that all objects with two blades are pairs of scissors. I might have, for example, a penknife with two blades.

In practice, this kind of deductive logic is often difficult to make use of, because in the real world it can be difficult to be absolutely certain about practically any statement beginning 'All'. A classic example is:

All swans are white.
This bird is black.
This bird is not a swan.

which would have been assumed be a true logical deduction until visitors to Australia discovered black swans. In reality, all we can often get away with logically is induction. We can't say that all swans are white, merely that all the swans we have observed so far are white, so it is reasonable to induce that all swans are white. As it turns out, they aren't.

Rather than deal with such fluffy concepts, Boole transformed logic into a pure symbolic form, assembling an equivalent to arithmetic that dealt with the logical states 'true' and 'false', which could easily be represented as 1 and 0. These would become the logical operators of computer gates, dealing with binary values. We can make use of this approach for the kind of text statements used above, or for totally generalised statements. So, for example, if statement A is true (planets travel in elliptical orbits, say) and statement B is true (the Earth is a planet) we can say that A AND B is true. But if we have a different statement B (the Earth is a grapefruit) that is false, then A AND B is false. But in both cases, A OR B is true.

Though he had no idea it would be the case, the exact mechanism that Boole developed to provide a symbolic representation of logic statements proved to be the central mechanism of computer operations – and would prove central to the mechanism of specifying exactly what it is that we are looking for when scanning a database or typing a phrase into a search engine such as Google.

Speaking to a search engine

In principle, using a search engine is very simple. You type some text into a box, click a button, and the search engine matches that text with its index, finally returning a list of appropriate matches on the screen. But how does the software decide what is an appropriate match for the information that you have typed?

In the early days of search engines,* this involved strict Boolean algebra. If, for example, you wanted to find information on a male person who was a politician, by typing 'male AND politician' you would force the search engine to only return information on male politicians (or, to be more precise, on web pages containing both the word 'male' and the word 'politician'). Similarly, typing in 'dogs OR cats' would find pages that referred to dogs, or cats, or both. In practice, modern search engines such as Google and Bing are more forgiving with the wording of search queries and will interpret what you type more flexibly, although there is still a mechanism available to force a true Boolean search.

However, despite the cleverness of these searches, they do rely on more precision than humans and human brains are capable of in their pattern-matching. They are inflexible in the sense that you can't ask for something like 'the railway place in that town with the tower' and necessarily succeed. What I was looking for by asking this in my fuzzy** way, was Blackpool station. In the UK, at least, Blackpool is known for its small-scale version of the Eiffel Tower – and I was having a senior moment and couldn't come up with the word 'station'. I tried this search on Google, which first came up with the Wikipedia entry on Alton Station (formerly known as Alton Towers Station). It was only after 29 results that there was a mention of Blackpool, and then it didn't refer

---

* We're going way back to the 1990s.

** 'Fuzzy' is the genuine technical term for this kind of vague data terminology.

to the station. (In a fortuitous coincidence, an earlier result came up with the station of my birth town of Rochdale.)

Let's take that kind of fuzzy search up a notch. Imagine you were looking for a specific building – but you can't remember exactly what it's called, or which city it's in. You might say to someone else, 'What's the name of that building, it's a bit like a partly open book in shape, very thin in profile, at the convergence of two streets. Maybe fifteen storeys high.' And (if you're lucky), they might respond, 'Do you mean the Flatiron Building in New York?' Although they have got better, search engines are less effective at such fuzzy queries* – they like the wording they are given to be precise.

In 2000, a few years after coming up with his first algorithm, Lov Grover produced a second quantum computing algorithm which can deal with fuzzy, uncertain, unstructured data. When asked about its use a few years ago, he gave the example of searching in a phone book, though now his example would be much more relevant to any online search. For example, you might be trying to get back in touch with someone whom you met recently. She's called Annie and has a common surname, but you can't remember what it is. It might be Smith – maybe you'd give that a 50 per cent chance. Or Brown, perhaps. Say 30 per cent chance. Or was it Jones? That's less likely, but maybe 20 per cent. She mentioned she could see the Shard in London from her flat, and the last three digits of her mobile number are the same as your cousin Peter's.

---

* For your entertainment, the first four results from Google on this search were a book called *Introduction to Computer Information Systems*; intriguingly, a blog article on 'Finding a book when you've forgotten its title'; an article on 'London's local high streets'; and a PDF file entitled 'Kevin Lynch'. Not quite what I was after.

It's a jumble of information which is highly unlikely to be 100 per cent accurate – but it's exactly the kind of messy tangle that our brains deal with all the time. For a computer it seems extremely challenging, but this is the kind of fuzziness that Grover's 2000 algorithm handles well, able to reach the most likely result exponentially faster than a conventional computer – which almost certainly couldn't cope with this degree of vagueness at the best of times.

The prime un-multiplier

The other well-known algorithm for quantum computers from the early days, predating Grover's by two years, was Peter Shor's factorisation algorithm. Like Grover, Shor was working at AT&T's Bell Labs and in 1994 he showed how a quantum computer would be able to work out which two prime numbers had been multiplied together to produce a particular outcome. As we have seen, this would be a hugely significant outcome if it were practical to achieve with very large numbers, as it is extremely hard to extricate the two primes from the multiplication, a fact relied on by the current encryption used on the internet. The encryption technique typically uses variants of an algorithm called RSA, named after its inventors, Ronald Rivest, Adi Shamir and Leonard Adleman.

The three computer scientists developed RSA at the Massachusetts Institute of Technology (MIT) in 1977, though in fairness it should be mentioned that the mechanism was actually developed three years before this at the UK's intelligence centre, GCHQ, by Clifford Cocks. Unfortunately for Cocks, the algorithm was considered useful for national security by his superiors, so it was kept secret until the RSA trio had made their approach public, by which time Cocks missed out on any potential patent benefits.

Working through the entire mechanism of RSA is a little fiddly, but at its heart it depends on the person who is going to receive an encrypted message taking two extremely large prime numbers* and multiplying them together. The resultant huge number is given out freely, along with some other information required to work the algorithm, but only the recipient knows the values of the two primes that were multiplied together. The clever thing about this public/private key approach is that you can give the key to encrypt the information to anyone, but the different key required to decrypt it, is kept secret. The RSA algorithm uses the extremely large multiple to encipher the message in a way that it can only be deciphered if you are aware of the two primes that were multiplied together, which are only known by the recipient of the enciphered message.

Finding the prime factors of a number might not seem too much of a challenge. For example, if the large number were 91, it wouldn't take too long to work out that the two primes that had been multiplied together were 7 and 13. But with a number that is truly huge, working out the prime factors could take the best computers centuries. At the moment, the most common size for the large number is 2,048 bits of binary – over 600 digits in decimal, though increasingly 4,096-bit public keys are being used and there is no theoretical upper limit. You can always load more bits onto a key, though as the key gets bigger, the time taken to encrypt and decrypt goes up – so there is a balance to be struck.

---

* In case the concept is hazy, prime numbers are whole numbers larger than one which can only be divided (without a remainder) by themselves or 1. The first few primes are 2, 3, 5, 7, 11, 13, 17 ...

With Shor's algorithm, however, the size of keys most typically in use at present can be broken down into their factors on a timescale that potentially could put some of the encryption systems of the internet under threat. It's not all bad, of course – as we have seen (page 71), quantum technology does make totally unbreakable encryption possible, and there are alternative conventional techniques which become more practical as new computers get faster every year. However, there is no doubt that as and when fully functional quantum computers become readily available, some aspects of internet security will need a significant overhaul.

This may all seem speculative at the moment; however, the documents leaked by Edward Snowden back in 2014 showed that the US National Security Agency was running a secret programme specifically to develop a quantum computer that could break internet security. It's not just criminals who have an interest in cracking internet encryption, as many messaging systems encrypt information as a matter of course, and have been used by terrorists and others whom the state has an interest in monitoring. However, there is no evidence that the NSA has been any more successful than the many more public attempts around the world to construct large-scale quantum computers – and the leaked documents seemed to suggest that, if anything, the agency was a little behind the leading laboratories.

This requirement to take encryption beyond our current capabilities is something that has been common knowledge in the field now for over 20 years, and as a result considerable effort has been put into ensuring that quantum-proof encryption is available without having to resort to as-yet-still-niche quantum encryption. There are existing techniques going beyond RSA, such as AES (Advanced Encryption Standard) that could be made too hard to break in any sensible timescale. There are also new 21st-century encryption algorithms, such as some based on mathematical structures known as lattices, which provide encryption that it is thought will not be susceptible to breaking by quantum algorithms.

For that matter, Shor's algorithm is not all about cracking codes, of course. Breaking large numbers down to their factors can be useful in a wide range of mathematical applications, as can Grover's search algorithms. One potential use for the latter is the so-called travelling salesman problem. This is the task that a satellite navigation system or apps like Google Maps have to perform when they pick the best route from A to B on the roads.

This is another of those apparently simple problems like factoring that in practice is pretty well impossible for conventional computers to crack once the problem gets up to a realistic scale: as the number of road junctions goes up, the complexity of the problem rapidly gets out of hand. In practice, satnav software solves the problem by approximation. It doesn't necessarily find the best possible route for you, just the best of the ones it has managed to calculate. But using a quantum algorithm, it should be possible to get to the best solution far more often than is the case at the moment.

Gambling with quanta

It would be tedious to work through all the current quantum algorithms, especially as some are limited to very obscure mathematical manipulations. However, one recently developed quantum algorithm is a good example of the potential available, as it builds on a widely used existing technique we have already met: the Monte Carlo method. This was developed during the Second World War to help predict the behaviour of neutrons in a nuclear reaction.

As a quantum phenomenon, the interaction of neutrons with other quantum particles involves probabilities rather than certainties. The idea behind the method was that in trying to predict something where you had incomplete information but knew some probabilities it would be possible to repeatedly run an algorithm with a random value underlying it – rather like repeatedly spinning a roulette wheel – and by doing this, build up an accurate prediction of the required value. It would never be perfect, but with sufficient simulation runs it should be good enough. The idea was first devised by physicist Stanislaw Ulam and developed further by John von Neumann, whom we met in Chapter 2.

We can produce the effect of a simple Monte Carlo method prediction without the need of our own random number generator by making use of publicly available data that has a random component, thanks to the work of the UK National Lottery. Lottery draws use machines designed as much as possible to select balls randomly from a pool of the size required for the game, and the numbers drawn are recorded on the Lottery's website. Specifically, I am going to look at the main Lotto game.

As a thought experiment, I am going to imagine that for some reason I don't know how many main balls* are used in any particular Lotto game, but I am able to see how many times each of the first ten numbers has been drawn, and I do know the total number of main balls that have been drawn

* The Lotto game has a 'bonus ball' that is not part of the main draw. I am ignoring this.

in all the games for which I have statistics. I can then use the Monte Carlo method to estimate how many balls there are in total.*

If we look at the statistics from just one draw, it will tell us very little. In the most recent draw at the time of writing, of the five balls that were drawn,** none of the numbers between 1 and 10 were drawn. From this, I obviously know that there must be at least fifteen balls, because five numbers were drawn other than 1 to 10 – but there is nothing else useful that I can say. But let's look instead at every draw there has been, giving us a reasonable set of random draw numbers to work with.

With the size of data set available, each main ball has been drawn several hundred times. From the data below I can see that the average number of draws for each of the first ten balls was 291.5 and the total number of balls drawn was 14,518. If those figures were representative of the entire set of balls – and as the number of times the experiment is run goes up, the outcome should get closer to the truth – this implies that there were 50 balls in the draw. The actual number of balls in the draw was 49. So, the Monte Carlo method has enabled us to make a good approximation to the values despite being given incomplete information. If we look at the entire distribution of the numbers of draws per ball, rather than just the first ten, we can see why the outcome was a little high:

---

* Of course, I do actually know how many balls there are in a Lotto game, but that's not the point. And knowing the right answer is useful in being able to check my prediction.

** If you are familiar with the UK National Lottery and know that there are six main balls drawn in the Lotto game, bear with me.

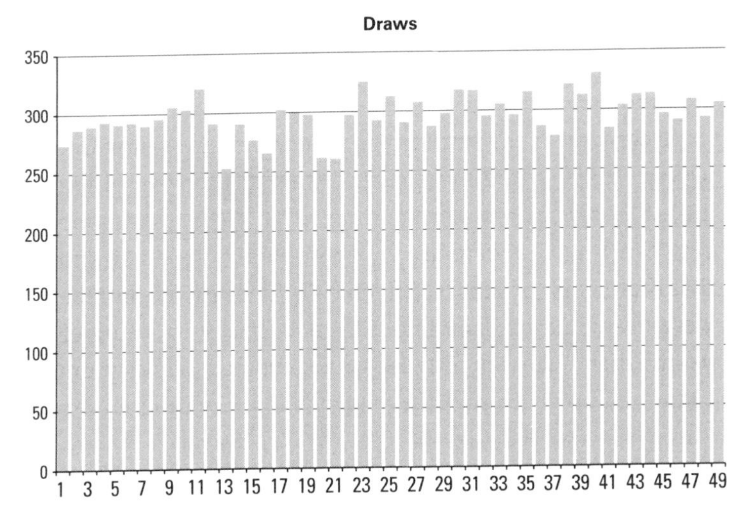

Number of times each main ball has been drawn in the UK National Lottery Lotto game.

Although the number of draws per main ball is settling down to a reasonably consistent level, there is still some variability, and, as it happens, the first ten balls aren't totally representative. They do, however, give us a good enough picture to get a useful feel for the actual value. Our estimate was only one out. (If you know the Lotto game, and realise there are actually 59 balls not 49, and that six numbers are drawn, not five, this is a relatively recent addition to the rules. In order to get a large number of runs of the game, it was better to stick to the 49-ball setup.)

The magic cocktail sticks

The lottery example is a touch artificial, in that it's not a particularly practical result. We did already know how many balls were involved. But it's useful because it is a genuine test of the Monte Carlo method using publicly available data. However, the same approach can be used to come up with a whole range of estimates, for example in predicting financial markets, weather forecasting, political forecasting, astrophysics, molecular modelling and many more applications. In fact, the approach was used long before it was formalised as the Monte Carlo method. The first recorded example was by Georges-Louis Leclerc, the Comte de Buffon, all the way back in 1777. He used the method to provide a way of calculating the value of the mathematical constant pi using only cocktail sticks, floorboards and a lot of patience.

Buffon, who was best known in his day as a naturalist, publishing a huge 36-volume encyclopedia of natural history, looked at what would happen if you repeatedly dropped a cocktail stick* shorter than the width of a floorboard onto the floor and counted how many times the stick crossed the cracks between floorboards. (You could use multiple sticks, but they tend to knock each other, and with sufficient of them in play, it can be difficult to see what's happening.) Buffon worked out that he could approximate to a value for pi (π) using the surprisingly simple formula:

$$\pi \sim \frac{2lt}{cw}$$

Here l is the length of the cocktail stick, w the width of a floorboard, t is the total number of sticks dropped and c is the number of sticks that crossed a crack. It might seem baffling

---

* In practice, cocktail sticks were probably not a thing in Buffon's time, and this method is known as Buffon's needle, suggesting he had another long pointy object in mind.

that pi, a value that we associate with the circumferences of circles, appears here, but bear in mind the chances of a stick crossing a crack will depend on the angle at which the stick lands – and once we get into angles, pi is likely to rear its head.

You can try out this experiment for real (draw parallel lines on a sheet of paper if you don't have visible floorboards), or, as I did, you can run a piece of software (see the Further Reading section for a handy online simulator). Just dropping ten sticks, I got five of them crossing a line, giving an estimate for pi of 2.5. I then dropped 100 more. This time, 45 crossed a line, making my estimate 3.055556. With 1,000 more, 469 of the sticks crossed the line, taking my estimate rather further away, at 2.958422. This is an important result, as it emphasises that the Monte Carlo method does have randomness at its heart. It doesn't unerringly home in on the right answer, getting more accurate with every cocktail stick dropped. Over time it will, statistically speaking, get more accurate – but in a random kind of way. You can imagine the value staggering randomly higher and lower around the actual value, but gradually being more likely to be close to it.

Adding 10,000 more – 4,363 crosses, my pi value got to 3.183016. In total, I dropped 101,110 sticks, got 40,165 crosses and an estimate of 3.146707 – not bad at all for a number which to those decimal places should be 3.141592. It takes a lot of runs of the algorithm to get reliably close – I had to run the simulation for over 500,000 stick drops before I got a result starting 3.141... – but it does very gradually home in. As with the Lotto balls, there are easier ways to calculate pi, but Buffon's method does illustrate the power of Monte Carlo methods to get a good approximation to an otherwise unknown value, provided we can put together an appropriate algorithm.

However, the example does also demonstrate a potential issue with the approach. Running this algorithm physically using cocktail sticks would take a very long time if we were to get a reasonable value. Remember I had to do over half a million (simulated) stick drops to get to 3.141... Performing such a number of drops is trivial with a computer – but this is a very simple formula. For a more sophisticated simulation, we could be pushing the limits of even the latest supercomputer. It's not for nothing that weather forecasters tend to buy the fastest machines available.

In 2015 Ashley Montanaro, from the University of Bristol, described a quantum algorithm which should require only the square root of the number of trials to reach the same level of accuracy. This would mean, for example, to get to my value of 3.141... I would only have to have undertaken around 700 virtual drops, rather than half a million. By 2019, this approach had been specifically tailored for pricing 'options' (these are financial contracts that give the purchaser the option of undertaking future deals). A research group at IBM, working with bankers JPMorgan Chase & Co., used a variant of the quantum algorithm which was shown to deliver the expected speed-up on pricing options using IBM's cloud-based quantum computing service (see next chapter for more on this).

A world of probability

When we run a conventional computer program, based on a conventional algorithm, we expect to always get the same outcome. Take as an example the pseudo-program to generate factorials shown on page 43. If we run that code multiple times, it will always produce exactly the same result. But when quantum functions are used there is always probability involved, and in practice a quantum algorithm will usually need to be run many times for us to have confidence in the outcome.

In this respect, getting a result from quantum algorithms could be more like the output of the software that modern meteorologists use to produce forecasts. Weather forecasting is very difficult to perform with any accuracy because the weather systems under study are chaotic in a mathematical sense: very small variations in starting conditions make big differences in how the weather will develop over time.*

Weather forecasting took a huge step forward towards the end of the twentieth century when forecasters stopped trying to predict exactly what would happen and instead ran their simulations many times with very small variations in the measured state of the weather systems at the moment that provided the starting point for the calculation. In such an 'ensemble' forecast, the meteorologists might run their program 100 times and discover that in a particular location, 60 of the runs predicted that there would be rain. They would then give the kind of forecast we see, of a 60 per cent chance of rain.

---

* The mathematical field of chaos theory began when a meteorologist, Edward Lorenz, re-ran an early computer forecasting program, but didn't realise he was using the rounded values of the parameters that were printed out with fewer decimal places than were used in the calculation, to save paper. The resultant forecast was totally different from the original one.

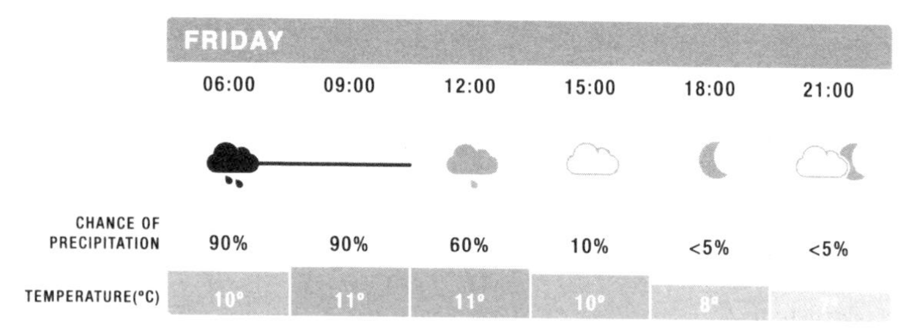

The 'chance of precipitation' in the forecast comes from running the software many times with slight variations in starting conditions and reports the frequency with which rain is predicted.

Similarly, when running a quantum algorithm, the results will typically be produced as probabilities of outcomes rather than certainties. Such algorithms can still be used to speed up processes, but this probabilistic nature has to be taken into account both in the way that the results are presented and in managing the expectations of those seeing those results.

The dangers of ignoring the probabilistic aspect of data are very clear whenever the media present numerical data to the public. At the time of writing the UK is in the middle of a general election campaign. News outlets provide a 'poll tracker' which keeps track of the various opinion polls. The lead one main party (Conservatives) has over the other (Labour) is generally reported in the media as 10 per cent, as the Conservatives are shown on 42 per cent and Labour on 32. And that's all that will generally be reported. But the poll tracker shows an extra piece of information (see below).

This is the 'likely range'. What this means is that given the accuracy of the polling, there is a high chance (the source website claims 90 per cent in this case) that the actual values are in the range shown. So, rather than showing a 10 per

Party	CON	LAB
Average (%)	42	32
Likely range	(38–46)	(28–36)

The percentage votes for the two main parties on the poll tracker.

cent lead for the Conservatives, what the polls actually show is a 90 per cent chance of them having a lead somewhere between 2 per cent and 18 per cent. This is a much more accurate picture of what the statistics tell us, but the media believe, rightly or wrongly, that the public can't cope with probabilities and ranges. Yet these are exactly the kind of results that will be produced by a quantum computer. If they are to become mainstream, this kind of uncertainty has to become more widely understood.

Going further

As noted previously, there are a number of other algorithms beyond search, prime factors and Monte Carlo methods, though many of these apply only to very specialist mathematical problems and may never have practical applications. As yet, though, the range available is relatively limited. Some of this may be due to the limitations that are imposed in dreaming up algorithms without an actual device to run them on, but it is entirely possible that the list will always be fairly short, as we shouldn't underestimate the difficulties of getting quantum algorithms that will run. However, Lov Grover commented in an interview with the author a while ago: 'Not everyone agrees with this, but I believe that there are many more quantum algorithms to be discovered.'

Even if Grover is right, quantum computers are never going to supplant conventional computers as general-purpose machines. They are always likely to be specialist in application. And, as we shall see, it is not easy to get quantum computers to work at all, let alone develop them into robust desktop devices like a familiar PC.

It's time we took a closer look at the hardware side of the quantum computing world.

QUANTUM HARDWARE 6

Richard Feynman remains, more than 30 years after his death, the physicists' physicist. A Nobel Prize winner for his work on quantum electrodynamics (the quantum physics of the interactions of light and matter), Feynman contributed to the Manhattan Project to develop the atomic bomb and was a key figure in the Challenger disaster enquiry, bringing to light the cause of the space shuttle explosion. He never wrote a book for the public, yet he is famous for the books that were developed from his lectures and conversations, ranging from enjoyable anecdotes to the surprise bestseller of his undergraduate physics lectures, collected in the so-called red books. With wide-ranging interests and a powerful imagination, Feynman arguably set in motion the hunt for a working quantum computer.

Feynman's quantum simulator

In 1981, Feynman gave a keynote speech at MIT with the title 'Simulating Physics with Computers', which was expanded into a paper called 'The Computer as a Physical System'. Simulation is a well-established role of computers, frequently used in physics and other fields. We all make use of the output of computer simulations in the form of weather forecasts, for example. A forecast based on a simulation cannot tell us what *will* happen, but instead, in the case of a weather forecast, is based on a series of models of the atmosphere, which are run multiple times with small variations in the way that they are set up.

These models are not physical objects in the sense of, say, a model car, but rather mathematical descriptions of reality, which simplify and approximate so that their users can take an extremely complex system and make it practical for it to be analysed. Similarly, such models are used to simulate the behaviour of everything from the interaction of quantum particles to the evolution of the entire universe. As we have discovered, the quantum world behaves fundamentally differently from the objects we are familiar with – and when simulating quantum events it is necessary to make approximations that take into account this strange behaviour.

However, Feynman suggested that there was a way to make a quantum simulation far better – why not make a computer that instead of using clear-cut 0 or 1 bits to simulate the probabilistic nature of quantum interactions such as superposition, used actual quantum particles as part of the computer, making it directly provide a more realistic model of quantum phenomena. Of course, strictly speaking, a conventional computer does use quantum particles in its calculations – specifically, electrons and sometimes photons – all the time. But they are not used in a way that derives any benefits from their quantum behaviour, except in specialist hardware such as flash memory.

At the heart of the problem that conventional computers have in simulating quantum behaviour is the difficulty of producing randomness.

Going random

As human beings, we struggle with randomness. A quick test. Which of these is a more likely sequence to be produced by a truly random device, repeatedly picking numbers between 0 and 9?:

6 2 3 1 7 2 9 4 or
6 6 6 6 6 6 6 6?

The first set of values 'feels' more random. The second has a clear pattern to it, and we associate randomness with a lack of pattern. In reality, each of these sequences is equally likely to be produced by a truly random process. Admittedly, there will be far more sequences that look like the top one than like the bottom one – but the specific sequences have the same chance of occurring (to be precise, a chance of 1 in 100 million).

When people are asked to write down a sequence of random values, they almost always put too few repeat values in it. In our minds, randomness should jump all over the place, and it feels wrong that repetitions should occur. But that's a problem with human perception. Computers don't have this problem. So surely there is no real problem in using a conventional computer to deal with the randomness inherent in quantum behaviour?

Superficially it appears that there is no problem. After all, anyone with access to a spreadsheet such as Excel can make use of a function called RAND that will generate a random number between 0 and 1. However, in claiming this, the spreadsheet software's writers are being a little economical with the truth. Conventional computers are deterministic. Given the same exact set of inputs they will always produce *exactly* the same output.* In most cases this is just as well. We wouldn't want a shop's till or a company's wage system to produce different values each time a calculation was run. But it does mean that the computers we use every day are incapable of producing random numbers.

In reality, the 'random' functions on a conventional computer, whether it's the laptop on your desk or a supercomputer in a university, are not random at all. If you start them the same way, they will always produce the same outcome. Instead of producing truly random numbers, they make use of what is known as pseudo-random number generators. These tend to involve a mechanism for starting with one number and generating a series of other numbers that jumps around in an *apparently* random fashion. But start the process with the same 'seed' (starting) number twice and it will produce exactly the same sequence.

To hide this deterministic nature, random number functions typically take as a seed the date and time at which the function is called, measured to a fraction of a second, so it

---

* Technically there are obscure circumstances, such as a bit being changed by a cosmic ray impact, where this is not the case, but on the whole a conventional computer running the same software with the same inputs will always produce the same results. In the real world, when things go wrong on a computer doing something familiar, it's because the user has done something different without realising. Or someone changed the program without telling anyone.

is very unlikely that they will be run twice with the same outcome. Surprisingly simple formulae can be used to generate a sequence of pseudo-random numbers that will prove good enough for most everyday requirements. For example, something like this:

$$\text{New Number} = (48271 \times \text{Previous Number}) \bmod 2147483647$$

This formula is a classic pseudo-random number generator known as a Lehmer generator, here using values that are widely employed to this day. The 'mod' is short for 'modulo', meaning that the new number is the remainder when the number in brackets is divided by 2147483647 (this, incidentally, is a prime number, which are often, though not always, selected for this role).

As the approach requires a previous number to get it started, this is where the seed value comes in, which might be, for example, the number of seconds since 1 January 1900 (at the moment I type this, that is 3,782,991,863 – though of course it will be more if you were to find out the value as you read this. The resultant new number will be somewhere between 0 and 2147483646 and jumps around satisfyingly. (Because we rarely want a number between 0 and 214783646, the function will usually divide the result by 214783646 to provide a more widely usable number between 0 and 1.) There are more sophisticated mechanisms for pseudo-random number generation, typically either variants of this approach or using cryptographic technology, but in essence the values labelled as random that are employed in conventional computers will only ever be pseudo-random.

Following ERNIE's lead

Pseudo-random values are usually fine for a quick spreadsheet calculation but can select some values too frequently, and there is always the risk that the same seed value will be used more than once, resulting in a repeat of the outcome. There are, however, ways around the limitations of pseudo-randomness if we look beyond conventional computers. Many lotteries use draw machines, such as those used to produce the values in the chart on page 87, which provide a good approximation to randomness as a result of being mathematically chaotic. This means that the output is still technically pseudo-random, but it is so sensitive to the way that the draw machine is started off that it is impossible to reproduce a run exactly, and the effect is to produce values that appear realistically random. But many UK readers will be familiar with another random number generator that employs true quantum randomness. It's called ERNIE.

ERNIE, standing for Electronic Random Number Indicator Equipment, is one of a series of machines of the same name dating back to 1957 that have been responsible for the monthly draw in a UK government saving scheme known as premium bonds. Unlike a conventional savings account, the interest from premium bonds is paid out as prize money to randomly selected winners, rather than being equally allocated to all investors. The original ERNIE machine, developed by Tommy Flowers and the team who built the Bletchley Park Colossus computer, got its random values from signal noise generated by neon tubes. It feels as if this is still pseudo-random, as the device made use of a similarly chaotic effect to that of the lottery draw machines, but in ERNIE's case, this was occurring at the electronic level in small variations in the physical conditions in the tube. This meant that, arguably, ERNIE's draws were genuinely random as the occurrences being measured involved quantum particles, each of which would have a distribution of locations, rather than fixed, predictable ones.

Later versions of ERNIE moved to basing its randomness on thermal noise in transistors – again, apparently pseudo-random, with small variations in physical conditions inside the device causing small variations in voltages, but nevertheless reliant on the behaviour of quantum particles and therefore probably truly random. Now we can do away with the 'probably' aspect. The most recent version, ERNIE 5, introduced in 2019, has an explicitly quantum random number generator, making direct use of the true randomness of quantum processes.

There are a number of units available to plug into conventional computers that generate true random values this way. Some make use of radioactive decay, the timing of which is random, while others, such as ERNIE, make use of the behaviour of photons, for example in travelling through a beam splitter (a quantum device to split a stream of photons, at its simplest a half-silvered mirror) or one of the effects that generates entangled photons known as parametric down conversion.

To show how these might be used, a beam splitter might send half of a beam of light in one direction and half in another. This feels perfectly natural when we think of light as a wave that can be split in two. But what if we send individual photons through the beam splitter? They can't be cut in half – that's the whole point of quantum particles. Instead, until they are detected as arriving somewhere, the photon is in a superposed state of going in both possible directions.

Eventually it will trigger a detector – and it will have gone one way or another. Which direction it took is genuinely random. It's this kind of optical randomness than lies behind ERNIE 5.

The power of analogue

Such true random number generators are doing on a small scale what Richard Feynman suggested in terms of computation. The best way to simulate a truly random quantum process is to use a quantum process. If, Feynman suggested, you had a computer which, instead of having conventional bits, operated on the potentially superposed states of quantum particles, you would have a computer that had the true ability to simulate a quantum system – because it was one.

Feynman was not thinking of a programmable computer in the traditional sense when he wrote his paper, but rather a dedicated simulator of quantum processes. This would have had some similarities to so-called analogue computers. Before digital computers became common, analogue computers were frequently found in universities, engineering establishments and businesses. Rather than being based on distinct numerical values, analogue computers made use of continuous values in physical objects and processes.

For example, the most popular simple analogue computer was the slide rule, used to make calculations by hand by sliding logarithmically calibrated bars* along each other.

---

* Logarithms work using the powers of numbers, which can be multiplied by adding those powers. So, for example, while I might struggle to multiply 16 by 64 in my head, 16 is 2^4 and 64 is 2^6, so to multiply them I add the powers to get 2^{10}, or 1,024. Slide rules

Other, more sophisticated devices might use flows of water or collections of electrical resistance to provide a calculation based on the physical outcome of the process. Analogue computers typically produce approximate rather than exact answers and tend to be designed for specific applications, rather than general-purpose use.

Some have suggested that once we have working quantum computers, we could turn this analogue approach on its head and use quantum devices to find solutions that cannot be achieved mathematically.

A physicist who believes that this is the case is Artur Ekert, professor of quantum physics at the University of Oxford's Mathematical Institute and one of those responsible for the development of quantum encryption (see page 71). Ekert describes a problem that is impossible to solve using mathematics, but that can be solved physically. Imagine there are two rooms, one with three old-fashioned incandescent light bulbs in it, the other with three switches. Each of the switches operates one of the light bulbs, but we don't know how they are wired up. If you are allowed only a single visit to each room, and it isn't possible to see into either room when you aren't in it, how would you find out which bulb was connected to each of the switches?

As a mathematical logic problem this is impossible to solve, because there are too many variable things for the number of times you can try something. But Ekert points out there is a *physical* solution to the problem. You go into the room with the light switches and turn switch 1 on for ten minutes, then turn it off. You turn switch 2 on and then

---

performed calculations like multiplication by turning numbers into the appropriate powers and adding them or subtracting them by moving one ruled bar against another.

leave the switch room and head into the bulb room. The lit bulb will correspond to switch 2, the warm unlit bulb will be linked to switch 1 and the cold unlit bulb is on switch 3.

Ekert suggests that there may be equivalent opportunities offered by the physical capabilities of a quantum computer that enable it to do things that a purely mathematical conventional computer can never achieve. Like the bulbs and switches problem, it would involve measurements of more than one physical property (in the case of the bulbs it was both light and heat) – but that's exactly what quantum computers make possible. Whether or not there will eventually be such remarkable abilities from quantum devices, within a couple of decades of Feynman's paper being published, scientists were beginning to try to work out how to use quantum particles in place of transistor-formed bits in computing. And part of the inspiration for doing this was the thought of a looming limit. The end of Moore's law.

How long can we carry on?

First things first – Moore's law, which predicts the rate of increase in the capacity of computers, is not a law. To be honest, the term 'law' has always been an odd one to use in science and technology. In the real world, a law is something that is devised by human beings and is specifically and tightly defined (as long as it's well written). It is literally a matter of black and white. It says what it says. On the other hand, such a legal law has no inherent control over nature. The people to whom the law applies have to agree to obey it, and there will always be those who don't, and who break the law, with variable consequences.

Physical laws, by contrast, aren't a matter of choice. You can't decide, for example, that you are going to break Newton's third law* tomorrow, however hard you try. But such natural laws are also a lot fuzzier, because there is no equivalent of a law book. Nature certainly *appears* to have laws – but they reside inside an unopenable black box. We can't see the laws in black and white, we can only attempt to deduce** what they are from observation of their effects. And our understanding of them is always subject to revision. So, for example, Newton's laws of motion are very good and useful in most ordinary circumstances, but they certainly aren't perfect reflections of nature, because the laws had to be modified by Einstein's relativity theories. With Einstein's work we now have a better match to reality, but there is no certainty that there aren't still gaps and flaws.

When we come to Moore's law, though, we aren't talking about either of these meanings of the world 'law'. Moore's law is really just a label given to an observed trend. The 'law' reflects the astonishing development of the power of the microprocessors at the heart of computers. It is based on a 1965 observation by Gordon Moore, then at Fairchild Semiconductor and later to co-found Intel, that the number of components on a chip was roughly doubling each year, and was expected to do so for the next decade. The predicted rate was tweaked down a few years later to predict a doubling

---

* In case your school physics is a little rusty, Newton's third law is the one that says every action has an equal and opposite reaction.

** As we have seen, in reality, what science usually does is induce (i.e. guess at the general based on specific observations) rather than deduce (draw a logically irrefutable conclusion from the evidence). But somehow 'deduce' works better in the English language. It's probably all Sherlock Holmes' fault.

of components around every two years, and this rough measure of the power of a chip has continued to follow the trend for another four decades.

Such consistent growth is remarkable, but we have to remind ourselves that this is no natural law. It's easy to assume that a trend like this will carry on indefinitely. But it's a mistake to do so. This is the kind of error that enthusiasts for technological growth often make. One example that has been made previously was to chart the unstoppable increase in speed of travel. For the majority of time that civilisations existed, we were limited to the maximum velocity we could achieve when running or when riding a quadruped steed. Less than 200 years ago in 1830, the first passenger steam railway opened, bringing speeds up significantly. With increasingly rapid succession, this was followed by cars, planes and jets. Then we headed into space and speeds took another huge jump.

What those who have used this example in the past seem to miss is that this was all achieved 50 years ago, and there have been no increases since. To put that into context, it has been about as long since Apollo (the fastest vehicle that humans have ever travelled in) was in action as it was between the Wright brothers' first flight and the first space launch. In fact, things are rather worse than the picture of acceleration that this progression paints. Space travel remains a very specialist capability. If you limit consideration to opportunities available to the general public, we have actually slowed down from Concorde's 2,000 kilometres per hour (1,300 miles per hour) in the 1970s to a mere 900 kilometres per hour (560 miles per hour), the typical cruising speed of a modern jet and the fastest a normal 21st-century individual can travel.

There is no reason why a trend should continue as it has before. The future is not always a straight-line continuation of the past. And the reality is that Moore's law is running out of headroom. Admittedly, people have been saying this since at least the 1990s, but there must come a point where conventional electronics is running out of ways to get around physical limitations. Perhaps the biggest hope at the moment comes with graphene and other ultra-thin materials, which can produce electronic devices significantly smaller than is possible with traditional silicon chips. However, even these workarounds will eventually run up against physical limits.

As we have seen (page 54), on a very small scale, quantum particles, such as the electrons in a wire or in a chip, do not behave as if they were solid particles passing through a tube. They are able to tunnel through barriers. As circuits get smaller, it becomes harder and harder to cram more into the same space without quantum effects disrupting operations. While it is likely that there will be a few more years of Moore-like growth, there is increasing recognition in the field of the need to move from what's referred to as 'more Moore' to 'beyond Moore' – looking for different ways to enhance the power of computers. And quantum computers are in the vanguard of this development, which is why there are many hundreds of teams around the world working on different aspects of quantum computing.

Enter the qubit

To see how a quantum computer can go beyond the capabilities of a conventional machine at the physical level, it's best to start with the basics. As we saw on page 20, at the heart of a current computer, whether it's a laptop, a desktop, a phone or the engine management system in a car, is a processor and its associated memory. These make use of bits, simple configurations of one or more transistors and other components which can hold an electrical charge or not, representing the binary values of 1 or 0. In a quantum computer, the bit is replaced by the qubit (pronounced 'cue-bit'): the unimaginative term for a quantum bit. These are typically individual quantum particles (or, perhaps more accurately, they are mechanisms for holding a quantum particle in place). And rather than storing information as an electrical charge, the qubit's value is determined by one of the particle's quantum properties, typically spin or polarisation.

As mentioned before (page 59), when measured in any particular direction, quantum spin can have only one of two values, up or down. So, at first sight, it resembles the 0 or 1 value of a conventional bit. However, because quantum particles can be in a state of superposition of those values, a qubit effectively stores both 0 and 1 at the same time. This feels as if it should double the capacity, which in itself has quite an impact if enough qubits are involved. For example, three conventional bits can store any one number between 0 and 7 (binary 000 to 111). However, three qubits can make use of all eight numbers simultaneously.

There's more to it than that, though, because a qubit's spin, for example, is not just the two values 'up and down'. It is more accurately 'up with probability x, and down with probability 1–x'. The two probabilities have to add up to 1 (or 100 per cent, if you prefer), but it doesn't have to be a straight 50:50 split. This is where things get particularly interesting, because, in effect, a qubit can represent any decimal value between 0 and 1, depending on the ratio of the up/

down split of probabilities. If, for example, you considered a 100 per cent probability of up as 0 and a 100 per cent probability of down as 1, then any intermediate value effectively represents a direction somewhere between up and down, which can represent an infinitely long decimal.

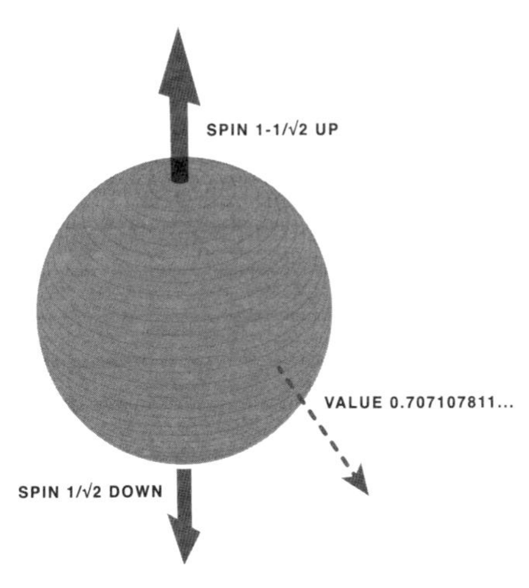

A qubit can represent any direction between up and down and hence any value between 0 and 1.

A useful analogy for the difference between a qubit and a conventional bit was provided by Tim Spiller, Professor of Quantum Information Technologies at the University of York: 'Picture it as a classical bit being only black or white, but a qubit having every colour you like.' Bear in mind we're not talking here about just the familiar rainbow colours, or the colour palette available to a computer, but a limitless range of possibilities.

And it gets better still. In reality, a qubit is even more impressive than we have suggested so far, because the qubit isn't a flat thing on a page as represented in the illustration above. The total state of a qubit usually requires two complex numbers to represent it in three dimensions. To get a clear picture of what's going on with these, we need first to be sure what complex numbers are.

It's getting imaginary

For centuries, mathematicians have had fun playing around with numbers that don't have any direct correspondence to objects we can experience in the physical world. Arguably, the first such numbers to be used were negative numbers. I can't show you –3 oranges or a negative amount of liquid in a bottle. But I can, at least, use the negative value indirectly. If I take away three oranges from the five you have to leave two (say), you could say that I have *given* you –3 oranges.* One of the most powerful such 'non-real' numbers are the imaginary numbers, which are based on the square root of –1.

In the arithmetic of real-world values, imaginary numbers don't exist. The square root of any particular number is the number which when multiplied by itself produces that number. The square root of 9, say, is 3 because 3 × 3 = 9. When you multiply a positive number by itself, you get a positive number. When you multiply a negative number by itself, you get a positive number. (So, the square root of 9 is also –3, because –3 × –3 = 9.) There is no obvious type of number that will produce a negative value when multiplied by itself. As a result of this limitation of reality, mathematicians simply define *i* as the number which when multiplied

* Though the chances are that you would insist I took three of your oranges.

by itself produces –1. Other imaginary numbers are shown using *i* and a multiplier, such as 3*i*, which would be the number that multiplied by itself produces –9. Mathematicians can do this because they control their numerical world.

Initially, the square root of a negative number was regarded as a useless novelty. But as mathematicians and scientists became more familiar with them, imaginary numbers proved extremely valuable, especially when combined with the usual 'real' numbers. This was because any point on a two-dimensional plane can be represented by a combination of a real number and an imaginary number, forming a so-called complex number. And the arithmetic of complex numbers is perfect for describing a two-dimensional real-world entity that evolves with time, such as a wave. There may not be imaginary numbers in nature, but they proved very effective in the mathematical world as a way to describe nature, so they came to be used extensively in physics and engineering.

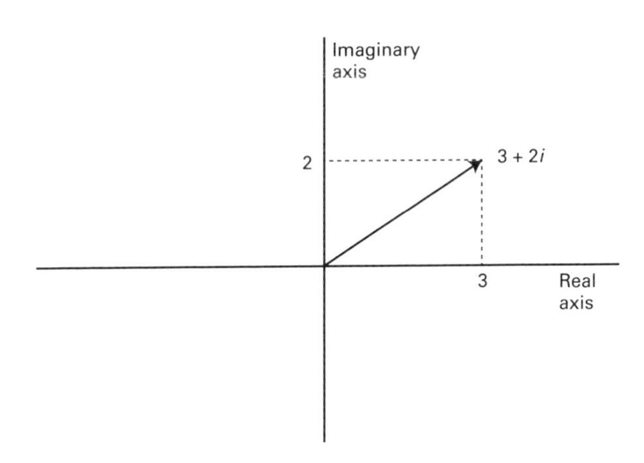

The complex number 3 + 2*i*, being used to represent a point on a two-dimensional plane.

As an alternative way of representing the value of a qubit we can think of it as being represented by two limited complex numbers: with a single complex number we can represent a direction in two dimensions, while the other makes it possible to extend this to describe a direction in three-dimensional space. The complex numbers are not free to take any values, because we are only interested in an overall direction: the squares of the two probability values that make up the real and imaginary components of the complex number must always equal 1. So the length of the arrow in the above diagram is always 1. This gives us a slightly more complex picture of a qubit, known as a Bloch sphere, after physicist Felix Bloch who devised it.

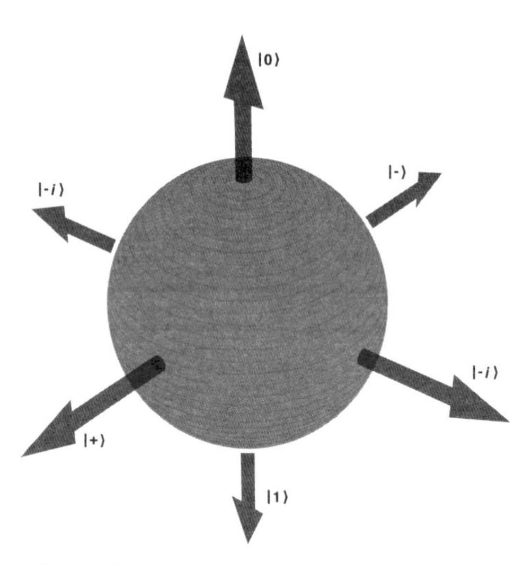

A Bloch sphere showing how the three axes represent the state of a qubit.

Each of the axes on a Bloch sphere represents a particular aspect of the quantum state of the qubit property (for example, spin) being used. We will come back in the next section to what the strange labels with the unusual straight and angled brackets around them mean.

Because of this three-dimensional storage, the capacity of even an apparently tiny quantum computer can be quite startling. The computer I am writing this on has 24 Gb of memory. That's 1.92×10^{11} bits. A fully functional quantum computer with just 100 accessible qubits could not only outclass my computer, but every supercomputer currently in existence added together. There are, however, two problems standing in the way of achieving this – making the qubits stable, and transferring information around (and in and out of) the computer.

Making qubits real

When working on quantum algorithms, we can talk about perfect qubits that perform as we want them to. But when building an actual physical quantum computer, we need to construct qubits that work. This means being able to get hold of a quantum particle, keep it in place and interact with it, making use of its superposed state without the qubit undergoing decoherence, the process whereby it interacts with its environment and loses its superposition. We could in principle use any quantum particle – a water molecule, say – but it seems obvious to stick with simple particles, which will presumably be easiest to deal with, and a significant amount of effort has been focused on two particles we've already met a number of times: photons and electrons.

Photons are plentiful, are now relatively easy to produce in a controlled way, and are the easiest particles to get into an entangled state, which is often necessary in a quantum computer. Photons are also relatively easy to keep in a superposed state, as they don't interact with each other at all, and are less likely to interact with their surroundings than a charged particle like an electron. It's not all good though. Light is not in the habit of hanging around until you need it. In fact, Einstein's special theory of relativity was built on the idea that light always has to travel at a fixed speed in any particular environment. So to pin down a photon is a tricky task.

In essence, the only way to trap a photon is to get it into a tiny reflective box. The photon will then bounce around inside. Even then, with conventional boxes, there is a tendency for the photon to undergo decoherence on reflection.* In 2019, researchers at the City University, New York found a new way to store single photons, holding them in a state that is particularly robust in the face of decoherence and capable of releasing them on demand. The photons are held in reflective cavities, which are partly open. This may seem to make it easy for them to escape, but it proved possible to use a form of interference to keep them in place.

Interference is a process undergone by waves. When waves are oscillating in the same direction, they reinforce each other, while those that oscillate in opposite directions cancel each other out. Here, it is the probability waves associated with the quantum particles that interfere with each other. This mechanism is theoretically so effective that it also

---

* Because we're taught it this way, we tend to think of reflection being a matter of photons of light bouncing off a mirror like balls off a wall. In reality, reflection usually involves the photon being absorbed by an atom in the mirror, then another photon being re-emitted. This makes the process much more dangerous for the possibility of decoherence.

stops photons getting into the cavity in the first place, but the researchers discovered that when two photons hit it at once, one is lost and the other is trapped as the system closes on it. According to one of the researchers, Michele Cotrufo, 'The stored photon has the potential to be preserved in the system indefinitely.'

When the cavity is hit with a second burst of photons, the original one is released, potentially making photons more feasible to use as qubits. This approach has yet to be tested experimentally, but is promising for the future.

Electrons also have the potential to provide good qubits. We know a lot about dealing with electrons – the ability to handle them is at the heart of electronics. Certainly, they are far easier to keep in place than photons, for example using a technology known as quantum dots. These are tiny pieces of semiconductor which act almost as if they were artificial atoms, trapping electrons in a way that the electron can be boosted into different energy states, just as they can with atoms (more on this in a moment).

However, in practice, it has proved surprisingly difficult for electrons to take on the qubit role. They were often the subjects of early experiments, as they work fine as individual qubits or even in pairs. But a quantum computer needs there to be interactions between a number of qubits within the computer to give it any practical usability. Getting an interacting structure of electron-based qubits has proved difficult to date. Even so, some work has been done on using electrons this way, employing microwaves to link them together.

The most-used qubits so far have been rather larger quantum particles: ions. An ion is just an atom which has gained or lost electrons, giving it an electrical charge. Because it is charged, it can be held in place using a so-called ion trap – a device that uses electrical and magnetic fields to float the ion in a vacuum, making sure that it does not come into contact with its surroundings and hence making it slower to undergo decoherence. The first ever quantum gate, dating back to 1995, was constructed using ions in traps* and since then a wide range of steps forward in the development of quantum computing have involved ions.

Rather than deal with spin, in an ion it is the energy levels of electrons that are used to form the qubits, as is the case for electrons in quantum dots. The idea of energy levels dates back to the very early days of quantum physics. As we saw on page 52, Niels Bohr developed a model of the atom where electrons could occupy different 'tracks' around the nucleus, but could not exist in between these levels. When absorbing the energy of a photon, an electron jumps up an energy level, and when it gives off a photon it drops down an energy level.

When ions are used as qubits, the property of the quantum particle that is put into a superposed state is the energy level of one of its electrons. As this is a quantum property, it is entirely possible, if it does not interact with its environment, for an electron to have probabilities of being in more than one state, and these different probabilities provide the superposition to enable the qubit to function.

Ions used in this way need to be cooled to near absolute zero to avoid the vibrations that arise from thermal energy, which could either disrupt the trap or produce decoherence. Temperature is a measure of the energy in the component atoms of whatever is being measured. That energy comes in

---

* Physicist David Wineland shared the 2012 Nobel Prize in Physics for this work.

the form of movement, as atoms fly around in a gas or liquid, or vibrate in a solid. There is also the energy of the electrons if they aren't at their lowest energy level around the atom.

As we cool an object down, the atoms have less and less energy. At some point they have to run out of energy entirely and do not have anywhere lower to go. This is the temperature known as absolute zero, which is –273.15°C (–460°F). Most quantum computers need to be lowered to temperatures near this. Such temperatures are usually measured in kelvin, units of the same size as degrees Celsius, but starting at absolute zero – so 0 K is the same as –273.15°C. Electrons can be used at a relatively toasty 1 K, but ions need to be cooled to around 0.002 K. Lasers are then used to interact with the ions.

Although the ions are held away from the wall of the trap electromagnetically, they could still be forced into decoherence if an atom from the air collides with them. As a result, as well as being supercooled, the ions also need to be in an extreme vacuum to minimise the chance of interference. A suitably low-pressure chamber can keep ions away from collisions for up to half an hour.

An example of one of the most advanced ion-based quantum computers to date is the IonQ system, which was developed in 2018. At the time of writing, the company has built three devices. These make use of ytterbium ions, which have a single positive charge after the removal, using lasers, of an electron. As a result there is a single electron on the outside shell of the atom, which provides the functional part of the qubit.

Each ion is held in a tiny space surrounded by 100 electrodes which hold the ion in place electromagnetically. At the time of writing, the company has run single-qubit quantum gates (more on quantum gates soon) based on 79 ions, and has linked up eleven ions as a multiple-qubit gate. Rather beautifully, to read a value off the ions they are hit with lasers – if the ion glows it represents a 1 and if it doesn't it's a 0.

Other possible qubit structures include Josephson junctions (see page 137), nitrogen-vacancy centre qubits which make use of defects in diamonds (page 131) and topological qubits. These are based on familiar particles such as electrons, but make use of a so-called topological effect, where the particles are able to resist decoherence because they effectively double up on their values, simultaneously holding the qubit value in two different ways.

Quantum gates

Just as a conventional computer has circuits that act as logical gates, performing operations such as NOT, AND, and OR, so the quantum computer has quantum gates, which inevitably are more complex than a conventional computer's gates because they are operating on multiple probabilities at a time rather than just 0 and 1. In one respect, though, quantum gates are simpler, in that (apart from some special operations such as measurement), they are all reversible.

A gate is reversible if going through it twice gets you back to where you first started. The only one of the conventional gates that is reversible is NOT. If you remember, this turns 0 into 1 and 1 into 0. So, going through two NOT gates brings you back to the initial value. But there is no equivalent ability for the other gates.

The most basic quantum gate, equivalent to the NOT gate is the X gate. Where the NOT gate swaps the values

0 and 1, the X gate swaps the probabilities of the values, known as their states. The states of the various properties of a qubit are represented using a general quantum notation that goes back to the work of the English physicist Paul Dirac in the 1930s. If we have a measurement of a property such as spin, that comes out up or down when measured, which we will represent as 0 for up and 1 for down, then the two states are denoted as $|0\rangle$ and $|1\rangle$.

This notation $|?\rangle$ is known as a ket, and has an equivalent written as $\langle ?|$ that is called a bra. Put together as $\langle ?|?\rangle$ they make up a bra-ket – a bracket.* This weak humour that Dirac came up with did not have the connotations it does now, as a 'bra' in the sense of an item of underwear got the name around 1936, after Dirac introduced the terminology – but it has stuck despite the traditional reaction of undergraduates when first introduced to it.

If a qubit is in a superposed state, with different probabilities for being measured as up (0) and down (1), the combined state is described as $a|0\rangle + b|1\rangle$ where the sum of the squares of a and b are 1. These squares of these values are the actual probabilities, which can range from 0 as no chance to 1 as certainty. So, for example, when there is a 40 per cent chance of up (0) and 60 per cent chance of down (1), a^2 is 0.4 and b^2 is 0.6. So, the combined state is $0.6325|0\rangle + 0.7746|1\rangle$.

Once we have this notation, we can see that the result of putting a qubit through an X gate is to change the state of the qubit from $a|0\rangle + b|1\rangle$ to $b|0\rangle + a|1\rangle$ – it swaps the

* For those with a mathematical background, a ket is a column vector and a bra is a row vector, while a bra-ket is the inner product of the two vectors.

probabilities. If we revisit the Bloch sphere that was shown above with a little more detail:

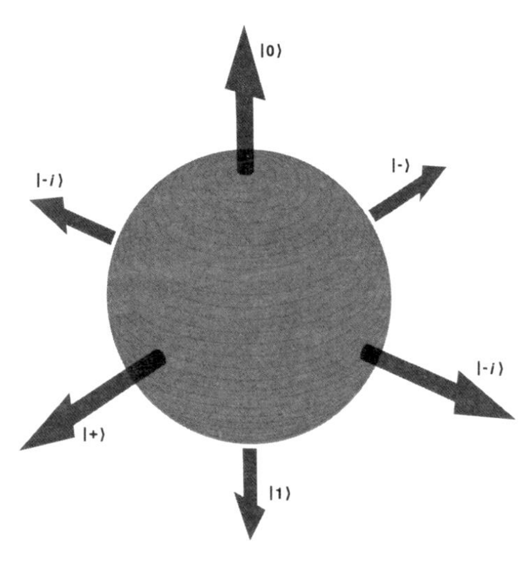

Bloch sphere with state values.

we can see that the X gate swaps along a single dimension that we have represented here as up and down. Two other quantum gates, imaginatively known as the Y and Z gates, swap probabilities along pairs of dimensions. So the Y gate swaps $|0\rangle$ and $|1\rangle$ as well as swapping $|+\rangle$ and $|-\rangle$, while the Z gate swaps $|+\rangle$ and $|-\rangle$ along with $|i\rangle$ and $|-i\rangle$.

One of the most frequently used quantum gates is the Hadamard gate, named after the French mathematician Jacques Hadamard. (This is not because of any involvement in quantum computing – Hadamard died in 1963 – but because the mathematical transformation of states involved correspond to a transformation of matrices that Hadamard described.) The gate has a somewhat convoluted mathematical function, but its effect is to put a qubit into a superposed state with equal probabilities for $|0\rangle$ and $|1\rangle$. Beyond this there are a significant number of other gates which impose different kinds of rotation on the Bloch sphere of a single qubit.

To date, a lot of work has been done on single qubits because they are easiest to set up. However, to do anything practical it is necessary to scale up. Just as in conventional computing, most gates operate on more than one bit at a time: a usable quantum computer must be able to work across multiple qubits. In the quantum domain, multiple-qubit operations are undertaken by entangling the particles which form the qubits and extended forms of the gates are available to apply operations to multiple entangled qubits, such as swapping the probabilities attached to properties of the linked qubits, or in the case of a special CNOT gate, undertaking the role of creating entanglement in the first place.

Fixing errors

The designers of any system handling data have to be aware of the potential for error. Where possible, at the very least, systems should be able to recognise that an error has occurred, and, if possible, fix it. For example, the ability to detect errors is often built into long numbers that are likely to be typed into a computer by human beings, as it's not unusual for us to mistype a digit or swap a couple of numbers as we go. A familiar example of this is the ISBN* used to identify books (found on the back, by the barcode), where the last digit is not part of the identification number, but rather a check digit. This 'fake' last digit also applies to an

---

* International Standard Book Number

even more familiar long number that most of us have to sometimes type or speak – a debit or credit card number.

The card numbers used by most banks are sixteen digits long, of which the first fifteen are the actual card number, while the last is a check digit which is calculated from the other numbers. A single digit can't check that everything about such a long number is correct, but the method used, called the Luhn algorithm, picks up both a single incorrectly typed digit and the swapping of two adjacent digits – a common error, particularly when typing quickly – unless those digits happen to be 0 and 9.

Imagine, for example, that my credit card number is 9192 8245 2272 1739 (as far as I'm aware this is a totally fictional number – at the time of writing, there aren't any card issuers with numbers that start 91). I start by chopping off the final digit and putting it to one side. Then I multiply every odd digit by 2. If the result is bigger than 9, I add the two digits of the value together – for example, if the result were 14 I would add 1 and 4 to give 5.* I then add together the fifteen values – all the new odd-digit values plus all the old even-digit values – to get a total. In the case of my imaginary credit card, that total was 71. Finally, I see how much I would have to add to that number to make it a multiple of 10 – in this case 9. So my sixteenth digit is, indeed, 9.

If, now, I accidentally mistype one of the numbers, the calculated value will no longer be 9. So, for example, if I make the second digit 7 instead of 1, the calculated value is 3. I check this against 9 and it's wrong – so the number was typed wrong and the computer will reject it. Similarly, if I

* The process of adding digits to form a single number is called taking a digital root.

swap the 8 in position 5 with the 2 in position 6, the calculated value is 6. Another fail. And so on.

This approach to errors detects them, but doesn't fix them. The system would reject the input and I would have to try again. That's fine for human interaction – but if it's an error within the system, it is better if the problem can be fixed. One way to provide a good chance of fixing an error is to repeat each digit a number of times. So, for example, if the number in question was the first four digits of my fictional credit card, instead of 9192, I could represent this as, say, 99999111119999922222. If, instead, the system detected 999991111199**7**9922222, it would seem likely that the error in the third digit was the 7 rather than the 9, so I could correct this back to 99999111119999922222.

This isn't a good way of checking for mistakes in human input on a keyboard, as the chances are, if your finger accidentally hit the wrong key you would type 99999111117777722222. It would also be extremely tedious to have to type each digit a number of times. However, if the system were trying to detect errors that have occurred due to some sporadic fault in a device, this would be a useful strategy.

Quantum correction

For a quantum computer, the biggest risk is decoherence. As we have already discovered, quantum particles lose their superposed state when they interact with their environment. This process is often described as undergoing measurement, but that implies a need for something active – for someone or some piece of equipment to undertake a measurement

– which is not the case. It can simply be a matter of a particle, for example, interacting with another particle (in the examples from pages 63–4, by passing through a polarising filter and interacting with the atom in that filter) and as a result ceasing to be in superposition.

To avoid such problems, a quantum computer will need to make use of more than one physical qubit in order to deal with a single qubit's worth of information (sometimes called a logical qubit). But adding in extra qubits itself increases the likelihood of decoherence occurring. While it is certainly possible to construct a fully fault-tolerant quantum computer, it is likely that there would need to be hundreds of physical qubits to provide one totally reliable logical qubit in practice. As this may not be realistic, it's likely we may need to allow a latitude for error that would not normally be considered in a conventional computer.

To make matters worse, though quantum teleportation allows us to transfer properties from one particle to another, it is impossible to use a simple copy approach as described with 99999111119999922222 above, which implies taking each value and reproducing it a number of times. Because of the no cloning theorem, a fundamental proof in quantum theory that it's impossible to make an identical copy of a quantum particle, we can't make multiple repeats of a qubit's state. Instead, quantum error correction has to deal with the problem in a different way.

A starting point is to not give the particle time to undergo decoherence, what might be called the 'hot potato' approach, passing the value from place to place before there's time for decoherence to occur. When the first experimental quantum computer rigs – little more than a collection of linked receptacles on a lab workbench – were established with just a handful of qubits, decoherence typically occurred within millionths of a second. The only way to achieve anything was to keep teleporting the property being used to another particle before the first had decohered. This is gradually becoming less of a problem, though we can never expect the same kind of robustness found in a conventional computer. In 2013, a team at the Canadian Simon Fraser University kept a set of qubits in a superposed state for around three hours at temperatures close to absolute zero, and for around 35 minutes if it was restored to room temperature after the qubits had been set up in the supercool state.

Like many of the early qubit experiments, this was not in any sense a working computer – there was no attempt to do anything with the qubits, though it is when a qubit encounters a quantum gate that is one of the most likely points for decoherence to occur. In practice, it has been estimated that around 90 per cent of the processing of a quantum computer will have to be dedicated to keeping on top of errors, making it workable at all.

Because of the inability to make copy qubits, quantum error correction generally relies on a mix of entanglement to link qubits together and the use of quantum gates, such as the Hadamard gate (see page 120). Back in 1995, Peter Shor came up with a structure of nine qubits which would automatically correct for either a swapping of $|0\rangle$ and $|1\rangle$ or an error that ended up changing the overall state from $a|0\rangle + b|1\rangle$ to $a|0\rangle - b|1\rangle$.

We might get better options developed in the future, but at the time of writing, around 1,000 physical qubits per logical qubit would be needed for full, failsafe error correction. At the moment, the biggest working devices have around 75 physical qubits.

Teleportation revisited

Now we've got an idea of the nature of quantum states and what a qubit is physically, we can revisit quantum teleportation to get a better idea of how this essential feature for communication between parts of a quantum computer is undertaken. As a reminder, what happens in teleportation is that the state of a quantum particle, such as its spin, is transferred from one particle to another particle, which could be located remotely. We don't find out what that state is, and the original particle is effectively scrambled in the process.

To make quantum teleportation happen needs a total of three quantum particles. One pair of these need to be already entangled. One of the entangled particles is required at each end of the teleportation process. These could be entangled particles created by a quantum gate, so adjacent to where they are required in a quantum computer, or they could be widely separated spatially, like the entangled photon pairs created by the Micius satellite (see page 68) and sent to ground stations 1,200 kilometres apart.

The third quantum particle, the source, which would usually be one that's being used as a qubit, is the particle that is in the state we want to transfer elsewhere – either to a different part of the quantum computer or to a totally different computer. If it's another computer, the mechanism is rather misleadingly referred to as using the 'quantum internet'. To make teleportation happen, the source particle is interacted with the nearby entangled particle, resulting in an instant change in the remote entangled particle. A measurement is made on both the nearby entangled and the source particle, which lose their superposition – the result of this measurement is transmitted as conventional data to the location of the remote particle, which, depending on the outcome, is put through a specific process. The remote entangled particle now ends up in the same state as the original source particle.

For example, if we measure the spin of the local entangled particle and the source particle vertically, then there are four possible outcomes: up+up, up+down, down+up and down+down. The measured outcome is sent conventionally to the location of the remote entangled particle, and that entangled particle is put through one of four possible processes, consisting of 'do nothing', and three different quantum gates. At the end of this process, the remote entangled particle now has had the state of the original source particle teleported to it.

Teleportation was first achieved in 1997, by Anton Zeilinger – a world expert on entanglement – in Vienna and Francesco De Martini and colleagues in Rome, using the polarisation of a photon, which was initially achieved locally on a desktop. However, Zeilinger is something of a showman, and by 2004 had achieved teleportation of polarisation from one side of the river Danube to the other. The entangled photons were transmitted through a fibre optic cable running through the sewer system under the city, while the information on measurements of local photons was beamed 600 metres across the river by microwave.

Getting information in and out

Entanglement and teleportation are fine for getting information between qubits, but at some point, we need to get information in and out of a quantum computer. Setting up values isn't too bad, as gates can be used to initialise qubits in various ways, but when we want to get a result out, we hit up against a particularly painful transition. Qubits are analogue, but as quantum objects, they only read out their states digitally.

Let's say, for example, we're using quantum spin for the qubit state and the qubit that holds the result of the calculation has a 30 per cent probability of being up and 70 per cent probability of being down. If we make a measurement of the spin in the up/down direction, we don't get 30 or 70. We just get either up or down. The outcome is quantised. Instead, to discover the value, we would have to run the program, say, a thousand times. Approximately 300 of these runs would come out up and around 700 down. After performing these runs we would have an approximation to our result. But like Buffon's needle (see page 88), we would never get an absolute, perfect value out*.

As much as possible, to avoid this, those designing quantum algorithms attempt to produce what are known in the trade as oracles. This is a reference to the Ancient Greek temple oracles. Although some of the oracles answered with detailed prophesies, others would only answer questions with 'yes' or 'no'. With careful design, as much as possible, problems tackled by quantum computers can be phrased in such a way that the only answer required is yes or no, and

---

* Note that the need to do multiple runs to get a result would be true even with perfect error correction: this is an inherent aspect of quantum measurement. If we didn't have perfect error correction, then even more runs would be required to minimise the impact of errors.

where, say a state measurement of $|0\rangle$ represents no and $|1\rangle$ represents yes. It's not ideal, but it's the reality of dealing with tricky quantum states.

Steps to a real quantum computer

As yet, quantum computers are mostly still at the laboratory worktop stage. At first sight, it might seem that the most significant step to take to make quantum computing more usable in the commercial world would be to move away from the need to chill them close to absolute zero. Clearly devices requiring super-cooled environments are never going to be available for sale in the local computer store – they will always be specialist.

However, the computer industry has a long history of running machines that require specialist environments. Until personal computers such as Apple and Commodore machines came along, the majority of computers had to be housed in specialist machine rooms – and that's still the case for the behind-the-scenes hardware behind, for example, a search engine. In these cases, it's less of a concern that specialist environments are needed, and, as we shall see, it's quite possible that most quantum computing facilities will be provided as cloud servers.

It's also true that, if necessary, equipment that needs extreme cooling can move outside of laboratories. MRI scanners, for example, present in most hospitals, make use of superconducting magnets, requiring specialist cooling. And the technology available to cool qubits is getting more sophisticated with time. It isn't necessary, for example, to cool the entire surroundings of the particle being used as a qubit. When a quantum particle absorbs and gives off light, for example, it can change its momentum. Specialist lasers can be used to effectively damp down the movement of atoms and ions this way, cooling the individual quantum particle and making it less likely to undergo decoherence without conventional supercooling refrigeration equipment.

Bearing in mind all the provisos we have about the limitations of current qubits and their assembly into working computers – and the limited error checking available – there have by now been a good number of experimental 'one-shot' quantum computations made: not employing a reliable, flexible quantum computer, but a setup in a laboratory where a quantum algorithm was successfully run at least once on a minimalist set of qubits.

For example, the first time Shor's algorithm for finding prime factors was successfully run was in 2001, when a team at IBM's Almaden Research Center led by Isaac Chuang made use of 10^{18} specially devised fluorocarbon molecules acting collectively. The seven atoms in each molecule provided the equivalent of seven qubits. These molecules were specially engineered so that the interaction between the atoms acted like quantum gates, while operating on an entangled cloud of atoms simultaneously made it possible to have sufficient results to overwhelm the errors and get a reasonable probabilistic result out. (Such an approach would not be practical for more complex calculations.)

Rather than work with electron energy levels as in the ion example (see page 115), the system made use of the quantum spin of the nuclei of the atoms. The outcome was detected using nuclear magnetic resonance, the phenomenon employed in MRI scanners. Inside an MRI scanner, powerful magnetic fields are used to flip the spin direction of hydrogen nuclei in water molecules of the test subject. When the magnetic field is removed, the nuclei flip back, sending out radio frequency photons. In effect, the water molecules in the subject are turned into tiny radio transmitters. The same principle was used in the quantum computer to interact with the nuclear spins in the atoms of the special fluorocarbon molecules.

As a result of this considerable effort, the IBM device was able to use Shor's algorithm to work out that the prime factors of 15 were 3 and 5. Getting these values isn't exactly hard to do. Not only could you do it in your head, as we have seen (page 30), Matt Parker has demonstrated a computer where the gates and data are constructed using 10,000 dominoes which are knocked over to perform a calculation of comparable complexity. However, the point is that 2001 attempt made use of Shor's algorithm and qubits to come to the result.

Another interesting example features an attempt at the Delft University of Technology in the Netherlands to move away from the need to cool qubits to near absolute zero by using diamonds as a host for qubits. These are the 'nitrogen vacancy centre qubits' mentioned on page 118: diamonds have a lattice structure which can hold impurities in the carbon lattice. The spin state of electrons on nitrogen atoms embedded in the lattice were used as qubits, which proved relatively stable at room temperature, because, in effect, the lattice of the diamond's carbon atoms acts as a shield to reduce the chances of an interaction with the environment that would cause decoherence.

It might seem as if diamonds would have become the environment of choice for qubits, because of this apparent stability and ability to work away from supercooled environments.* But in practice it's unlikely that diamonds will become a commonplace home for qubits. In part that's because it's non-trivial to find the qubits inside the diamond. Unlike a special trap that holds a single photon, electron, ion or molecule, the diamond has trillions upon trillions of atoms, among which the equipment has to home in on the few that are working qubits. Also, it's difficult to control the way that qubits scattered around a diamond are put into an entangled state, essential for many quantum computing operations.

Typically, the qubits in a diamond will either be isolated from each other, or the whole collection can be entangled with a blast of microwaves – but there is no obvious mechanism for exerting precise control. Worse still, each qubit in the structure will be unique, depending on the layout of the crystal around the nitrogen. In effect, each time a new diamond is selected, it will have to be individually set up for its structure, meaning that any kind of mass production is an impossibility. It's not just a matter of finding the potential qubits, but also the way that the system is wired up would have to be tailored to the layout of the diamond.

By 2019, the Delft team had a 10-qubit data store able to hold information for over 75 seconds, admittedly by resorting to not entirely room temperature conditions of 3.7 K. By this stage, they had moved from basing the qubits on electrons to primarily using carbon and nitrogen nuclear spins, a similar approach to the fluorocarbon molecule in the IBM experiment. From here they are intending to go on to including other materials such as silicon and silicon carbide.

* And also, perhaps, because using diamonds is rather cool.

It shouldn't have worked, though

Perhaps the strangest outcome of the relatively early piece of work by IBM on Schor's algorithm was the discovery that the IBM qubits shouldn't have worked at all. It turned out that the apparently successful quantum computer that discovered the factors of 15 ought to have been fatally flawed. The experimental setup couldn't have been safe from decoherence. The collection of fluorocarbon atoms was held at room temperature, and the entanglement between the computer's seven qubits would have collapsed long before a result could be obtained. Yet the experiment still worked.

It seems that it may be possible to make use of the messiness of a collection of less-than-pristine qubits and still do computation, a sort of make-do-with-second-best approach that goes under the name of discord. In a traditional quantum computer,* a quantum gate might take two or more pristine, entangled qubits as input and the result would be read off afterwards. But it was discovered that putting through the gate one traditional, cleanly set up qubit, carefully protected from interaction with its environment, and one qubit in a more 'normal' messy state that had been subject to measurement, would also make a quantum computation possible. The qubits couldn't be entangled, but there seemed to be enough of an interaction to allow the quantum calculation to proceed.

'Discord' is a measure of how much a system is influenced by observing it. A traditional classical system has zero discord, because we can look at it without changing the outcome. But any quantum system in a superposed or

* If this isn't an oxymoron.

entangled state has a positive discord reflecting the way that interaction with its environment risks decoherence. It seems that discord can give a degree of correlation, a kind of pseudo-entanglement between quantum particles that isn't so susceptible to collapse, linking together a mix of pure qubits and messy ones.

Even more so than with a perfectly set up quantum computer, the output of a discordant quantum computer involves results that are not exact but that have to be averaged across a number of runs. With sufficient repeats, though, the result seems to be reliable. What we have in a discordant computer is still a quantum device. It does require at least one pure qubit that is protected from decoherence, and although the rest of the qubits are in normal classical states, discord itself is a quantum effect. The whole process collapses and fails to work if that one pure qubit is allowed to go messy. But up until then, it is almost as if the addition of noise and disorder in the messiness of the rest of the qubits makes for a better, more stable quantum computer than one that is carefully protected from its environment – a decidedly hopeful thought for anyone attempting to build a commercial model and struggling with the menace of decoherence.

At the moment the approach is of limited use, because we only have the maths to be able to make use of very simple setups with discord linkages. As yet, the experimental physicists are waiting for the theoreticians to catch up. But there is a lot of promise there, and discord is being taken seriously. 2012 saw the first discord conference in Singapore, with over 70 researchers attending. This hybrid approach to quantum computing demonstrates in part why a technology that can sometimes seem impossibly far from practicality continues to be worked on all over the world.

Becoming reality

Beyond the basic testbed rigs of laboratories, any real quantum computer is likely to be mostly conventional – a large-scale, traditional computer that hands off some computations that are beyond it to a quantum module, which will need some kind of interface to the conventional digital world. Already IBM, for example, is offering the ability to interface with a very small quantum computer online – suggestive, perhaps, that in the future we aren't likely to see quantum computers on every desktop, but rather a conventional computer will hook via the cloud into a quantum computer in a specialist location, under the extreme physical conditions that most quantum computers require.

In a way, this vision of how quantum computers will work is rather similar to the way that modern astronomy is done. In the old days of astronomy, each university had its own observatory and an astronomer would use their own, local telescopes. Now most professional telescopes are at remote locations (or even in space). Instead of going to the telescope, the astronomer will hook up to it at a scheduled time from his or her office and make use of the telescope remotely until it's someone else's time slot. We may well see quantum computers develop in the same way.

In taking this approach, we are moving away from the notion of a quantum *computer* to quantum computation. In effect, the quantum part of the system becomes a secondary processor, just as, at the moment, we usually expect a computer to have a separate graphics processing unit to handle the specific calculations necessary to display graphics on a screen. The only difference is that rather than have the quantum processing unit built into the local machine, it is likely to be hosted remotely.

Achieving supremacy

A useful development in the history of quantum computing to get a feel for both the achievements in the field to date and the limitations attached to those achievements comes from Google's claim in September 2019 to have 'achieved quantum supremacy'. This is the rather self-important sounding terminology that describes having performed a calculation which would be infeasible using any current conventional computer.

The claim was given something of a sense of mystery when a paper on the achievement, published on the NASA website, was quickly removed before many people had read it. But the *Financial Times* newspaper had already downloaded a copy and reported on it. It seems that the Google team had pulled their draft from public view because they had found themselves a more prestigious home for the paper, which was published a few weeks later in the journal *Nature* as 'Quantum Supremacy Using a Programmable Superconducting Processor'.

The Google team had originally tried to use 72 qubits, but found this too difficult to control and, because of the time wasted on it, had to let an original prediction of achieving supremacy by 2017 go past. They cut down to a 53-qubit processor and made repeated experiments to be able to read off results, claiming that their processor, named Sycamore 'takes about 200 seconds to sample one instance of a quantum circuit a million times – our benchmarks currently indicate that the equivalent task for a state-of-the-art classical supercomputer would take approximately 10,000 years'.

That's quite a claim: but does it hold up to scrutiny? The task that the computer undertook was proving whether the numbers it had generated as random were indeed random – it was an introspective algorithm. The device was able to prove that the apparently random stream had an underlying pattern, where the conventional supercomputer was unable to do this. The authors suggest that this approach could also be used in some optimisation, machine learning and materials science applications, though the setup is not sufficiently fault-tolerant to achieve anything useful with the better-known quantum algorithms.

The superconducting qubits used here were so-called 'transmon qubits', a technology developed at Yale University in 2007, which makes use of quantum structures known as Josephson junctions. These junctions are pairs of superconductors with barriers between them. A superconductor is a material that is a perfect conductor of electricity – it has no resistance at all, so a current could flow forever. This is a quantum effect that usually only becomes available when a material is very close to absolute zero.

In such an environment, pairs of electrons join together in the configuration known as Cooper pairs which act as if they are a single entity that can pass without resistance through the material. These Cooper pairs tunnel across the barrier in a Josephson junction, and it's these pairs that provide the transmon qubit. To achieve this required serious cooling of the device, rather than just of the qubits, in this case reaching temperatures below 20 millikelvin.

The Google team's boast of supremacy was not to go undisputed. IBM's head of research, Dario Gil, described Google's claim as 'just plain wrong'. This is because the Google device was specialist technology, only capable of undertaking very specific types of computation, rather than being a general-purpose quantum computer able to take on the whole range of quantum algorithms. A few weeks later, Gil's colleagues at IBM, Edwin Pednault, John Gunnels, and Jay Gambetta, went further and challenged the idea that supremacy had been achieved, requiring as it does the quantum computer to far exceed the speed of a conventional machine.

According to the IBM trio, rather than taking 'approximately 10,000 years' to complete such a task on a conventional supercomputer, they estimated that it would only take the Summit supercomputer at Oak Ridge National Laboratory in Tennessee 2.5 days to finish it. At the time of writing, this is the fastest computer in the world, currently running at speeds of 143.5 petaFLOPS and theoretically capable of running at 200 petaFLOPS,* though the Summit is expected to be outmatched by the next-generation Frontier supercomputer, currently under construction by Cray Inc., for Oak Ridge.

Admittedly, 2.5 days is still significantly longer than the quantum computer took to complete the task – so while the experiment demonstrates an *advantage* for the quantum device over a conventional supercomputer, it hardly represents indisputable supremacy. According to the IBM researchers, the error made at Google had been to vastly overestimate the time the supercomputer would take, because the Google team had not considered the benefits of the huge amount of hard disk storage available to the supercomputer.

---

* PetaFLOPS is a measure of computer processing speed. 'FLOPS' are 'floating point operations per second' – operations on a floating point (decimal) number, while the peta prefix indicates 10^{15} – a thousand trillion.

Whether or not you agree with the fine points of detail, the naysayers have a point: this is not a practically useful application: it is a task designed around the device, rather than a device that was built to complete a worthwhile role. And the chances are that the supercomputer comparison was flawed. Even so, this was still an impressive demonstration of the potential of true quantum computers should they be built, bearing in mind it only involved a 53-qubit machine.

The quantum Galton board

There are so many labs pushing at the limits of quantum computing to try to be the next greatest thing that it would be tedious to list them all, but it is worth mentioning one alternative approach that, like the Google attempt, does not provide a universal quantum computer. It is another specialist device that makes use of a quantum effect to produce an outcome that would take significantly longer to reproduce on a conventional computer, but employing a very different and more robust concept of qubits.

This experimental rig works on a process known as boson sampling. Bosons are one of the two classes of fundamental particles, the other class being fermions. Where fermions could be typified as matter particles (such as electrons and quarks), bosons tend to be involved with forces, the best-known such particle being the electromagnetic force carrier, the photon. Boson sampling has been described as being something like a Galton board (a device apparently known as a 'bean machine' in the US). In these contraptions, used to demonstrate a mathematical principle known as the central limit theorem* – or simply as a game of chance – a ball is released at the top of a vertical table and as it falls, it bounces off a set of symmetrically arranged pins to finally end up in one of a number of pockets at the bottom.**

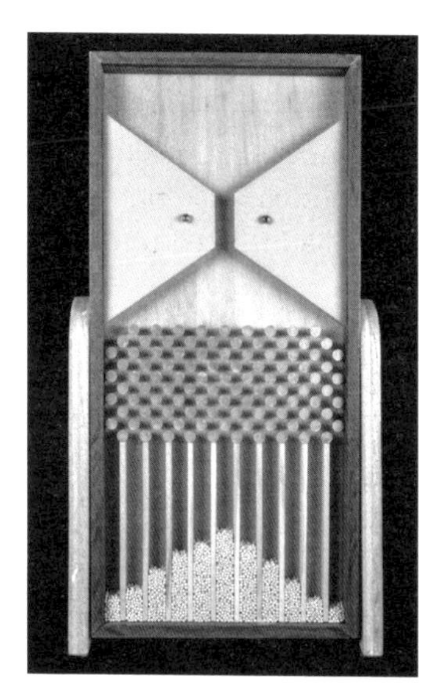

The near-random progress down a Galton board produces a normal distribution.

Matemateca (IME/USP)/Rodrigo Tetsuo Argenton

In boson sampling, photons are used rather than balls and the pins are replaced by beam splitters. As we discovered when looking at the ERNIE random number generator (page 101) these are devices such as half-silvered mirrors or double prisms which split a light ray's route and produce

* The central limit theorem effectively shows how a normal distribution (bell curve) builds up for appropriate random contributions.

** A similar device to a Galton board features in the game show on US and UK TV called *The Wall*.

superpositions when photons are passed through it one at a time. In November 2019, Jian-Wei Pan, Chao-Yang Lu and colleagues from the University of Science and Technology of China managed an operation where fourteen photons were detected at the end of the process. Because of the multiple interactions and superpositions, the boson sampling device effectively samples many different runs of the light-based Galton board simultaneously. The outcome is a hard one to compute with a conventional computer, so has the potential if more photons can be involved to demonstrate a limited form of supremacy.

It has proved difficult to scale up the experiment because the photons need to be both individual and produced at the same moment in time. In the best run, fourteen out of twenty photons made it through. The experimenters hope to get the experiment up to between 30 and 50 photons in a year, which should allow for sufficient complexity in the calculation that it would be able to demonstrate quantum supremacy, despite the single, again not particularly useful, application of the device.

You *can* buy a quantum computer ... sort of

It may seem a little strange to discover how far quantum computers seem to be from a commercial venture when you can go to a company and buy one now (at least, you can if you have a few million dollars to spare). The company in question is D-Wave Systems, which launched its initial prototype way back (in quantum computing terms) in 2007. This Canadian company came out with its first commercial product, the D-Wave One – with 128 qubits, priced at

$10 million – in 2011. Since then, it has launched a number of products, most recently at the time of writing the D-Wave 2000Q, with 2,048 qubits. D-Wave Systems expects to have a device based on its experimental Pegasus chip with 'more than 5,000 qubits' at some point in 2020. Not surprisingly, given the crucial importance of the potential search benefits of quantum computing, Google has been highly involved with D-Wave developments.

A D-Wave product looks like a typical supercomputer – a large, shiny commercial box sitting in a controlled environment rather than the go-anywhere laptop or phone, but it is still a packaged product unlike the one-off, lab assemblies that make up all the other quantum computers, operating (when they do so) with a fraction of the number of working qubits. The obvious question is why all this experimental work is going on if the quantum computer problem has been cracked by D-Wave. The answer to this is that, like Google's chip and the boson sampling devices, D-Wave is only a sort-of quantum computer.

Let's be more precise. D-Wave's products *are* quantum computers, but like those experimental examples, they make use of a very specific quantum process which doesn't provide the same flexibility and openness as a conventional quantum computer and that is unable to run the straightforward quantum algorithms we have already met. The main technology used by D-Wave's machines (there have been some variants) is an adiabatic quantum computer. But the good news is that it does seem to be significantly more capable than the other, more experimental non-general quantum computers.

The word 'adiabatic', from the Greek, roughly means 'not to be passed through'. It is commonly used in thermodynamics, the physics of heat, referring to a curve that displays the way a gas's pressure and volume change if there is no transfer of heat involved, or to a process that doesn't involve heat entering or leaving a system. In the D-Wave computers this indicates that instead of having quantum gates for its processing, it uses what is effectively a totally analogue computing approach, dependent on a process called quantum annealing – something, in fact, far closer to Richard Feynman's original vision than are most quantum computers.

'Annealing' generally refers to heat-treating a material – changing its properties by exposing it to heat. In the D-Wave it refers to using the lowest energy state of qubits to find a solution. The computer is first set up in such a way that the solution being looked for would be represented by qubits reaching their lowest possible energy states. Quantum effects enable the computer to effectively tunnel through barriers to get to low-level states that wouldn't otherwise be discovered. This is a very different process from the relatively traditional programming of conventional quantum algorithms.

One of the early demonstrations of an experimental quantum annealing processor with just four qubits managed to solve the problem of finding the prime factors of 143. (It's not RSA encryption levels – the primes are 13 and 11.) This wasn't done particularly speedily – a phone's processor could have achieved the outcome far faster – but managing it at all with just four qubits was impressive. Being a quantum annealing processor, though, it wasn't possible to use the ultra-fast Shor algorithm for this task.

It is certainly true that D-Wave computers are significantly faster than some conventional computers at specifically selected tasks. By the mid-2010s, D-Wave were pointing out that their computers could solve some problems 3,600 times faster than conventional software on digital computers. While true, the claim appears to be a little extravagant – the comparison was of a quantum annealing algorithm specially tuned for the specific purpose and no other, run on a multi-million dollar device, as against a piece of general-purpose software on a PC.

At the time of writing, D-Wave are claiming that there are now 'over 150 early applications' running on D-Wave, though it's not clear how many of these are faster than existing conventional applications or are capable of achieving anything practical. A fair number of the applications were in the area of optimisation, which often involves mathematical processes such as linear programming, which bears a structural similarity to the annealing method. It might be used, for example, to hunt for the lowest cost to perform a task, just as the annealing qubits hunt for the lowest energy state. Other suggested applications involved image recognition and simulations, again processes where the analogue nature of D-Wave calculations seem to be particularly appropriate.

There seems no doubt that D-Wave is here to stay, but in a sense, it is a distraction from 'real', more general quantum computing efforts.

More to come

Perhaps of all the topics covered in this 'Hot Science' series of books, quantum computing is most at a cusp. We have now had around 20 years of development of algorithms and experimental rigs – but is it ever going to become a significant force in computing?

TO INFINITY AND BEYOND 7

There is little doubt of the promise of quantum computing – yet, equally, the challenges should not be underestimated. We can already many see small-scale devices in laboratories, but they all are susceptible to significant errors because we simply don't have sufficient physical qubits available at once to make a device that is error-proof.

Are we there yet?

In the end, there is no point having an end product that gives no advantage over a conventional computer. It's fine for those building experimental rigs to get a better feel for how such quantum devices work. There will always be intellectual challenges around the delicate balancing act that is required to get qubits operating. But for quantum computing to be a significant resource in the real world, there is a need for a 'quantum advantage' – being able to do something that isn't currently possible in a reasonable timescale. Although we have seen Google's claim of quantum supremacy for their specialist device, this clearly is more of a publicity stunt than a significant quantum computing application in its own right.

The developments to date are hopeful signs, but there is clearly a long way to go. Physicist John Preskill, a frequent commentator on the field from the California Institute of Technology, Pasadena, has suggested that we are looking at something like a 30-year time horizon to reach full-scale machines capable of running general quantum computational algorithms. Some suggest that this is unnecessarily pessimistic – we have certainly made huge strides in the last 20 years – but others point out the parallel with the equally tricky task of developing nuclear fusion reactors. Experts have considered these to be around 50 years away from practicality for at least 50 years, and still are expecting such a wait before they are commercially viable.

Perhaps a closer parallel might be drawn with the development of artificial intelligence (AI). After huge initial enthusiasm in the 1960s and 70s, AI underwent a period that has been referred to as the 'AI winter'. The early enthusiasts vastly overpromised what was possible in a reasonable timescale and attempted to replicate the general intelligence of humans – something similar to the operation of a brain – rather than picking off specific, well-targeted applications.

In its revival since the 1990s, AI has picked out specific areas where there is some hope of considerable success. Although AI's abilities still tend to be oversold – for example with the idea that autonomous, self-driving cars are almost here, when they are very unlikely to be common on ordinary roads before the 2030s – there is no doubt that artificial intelligence is now achieving real and sometimes valuable outcomes, even though we have learned to be wary of its ability to confuse causality and correlation and to produce biased or unexplainable results.

Similarly, quantum computers have been repeatedly oversold, with tech firms underemphasising the difficulty of achieving error-free output, of constructing sufficient qubits or of keeping the computer running for a useful timescale. However, the small steps forward that we have seen are encouraging for the future, especially in the form of quantum computing as a cloud-based add-on for conventional hardware.

Baby steps

What we can see in the many experiments currently underway is a gradual erosion of the problems standing in the way of functioning quantum computers. For example, as we've seen above (page 115), ions are often used as experimental qubits. At the moment, quantum computers based on ions tend to be slow and limited in size, because the wiring and lasers used to interact with them take up too much room. But in 2019 a group at Oxford University managed to get around this particular issue.

Christopher Ballance and his colleagues produced short-duration laser pulses, which passed through a beam splitter before each of the halves of the beam was used to pump up the energy in a strontium ion. In this scenario, as the excited electrons in the ions lose energy, they get into a superposed state of two possible energy levels, as a result of which the photons they emit are entangled with the ions. The photons themselves are then entangled using a beam splitter, resulting in the ions becoming entangled with each other. Unlike other experiments, the ion traps are well separated – around five metres apart – so there is no scaling restriction.

To get a feel for the rate of advance resulting from this approach, back in 2007, it typically took around sixteen minutes to entangle a single pair of ions. By 2014, around five pairs of ions could be entangled per second. With this new approach, 182 pairs were being entangled each second. Although more is needed, this is the kind of dramatic scaling-up that is necessary to take quantum computing from the lab to practicality. This technique, for example, is expected to be able to get beyond 1,000 ion pairs entangled each second, and quite possibly 10,000.

Another possibility that could transform the industry is a move to silicon. As we have seen, at the moment most qubits are standalone physical objects, sometimes quite sizable where vacuum chambers etc. are required. Clearly it would be ideal if we could make use of our existing widespread expertise in production of incredibly complex silicon chips to make use of some form of qubit on a chip, based on existing construction techniques.

This concept has come up regularly and goes back at least to the late 1990s, when Bruce Kane at the University of New South Wales, Australia, came up with the idea of embedding phosphorus atoms in a silicon wafer, using the nuclear spins of the atoms as has been done in the diamond structures (see page 131), using nuclear magnetic resonance to interact with them.

Such a silicon-based quantum chip would have the potential to host sufficient qubits to enable large-scale error correction, perhaps holding between a million and a hundred million qubits. To avoid thermal noise, the system would still have to be cooled to close to absolute zero, but would be a lot more compact than many current arrays of qubits.

To date there have been experimental rigs using this technology for single and dual qubits – rather a way to go to reach the millions – and as other qubit approaches have made clear, it is the effective scaling that will be the essential test. From the experiments, it seems likely that quantum dots (see page 115) embedded in the silicon are likely to work better than phosphorus atoms.

There's plenty more to come

While the quantum computing revolution will probably take at least a decade or two to mature sufficiently to have the same kind of impact as AI is beginning to have now, the technology is moving in the right direction.

When Richard Feynman talked about quantum computers back in 1981, the field was little more than science fiction. We have been incredibly successful at taming quantum technology in the form of solid-state electronics, but the quantum computer might well have seemed one step too far. Now, though, we can expect a constant flow of breakthroughs. Qubits will never take over the world – but they are about to make their mark.

The quantum computing revolution is a slow climb, rather than a dramatic transformation. But it is underway.

FURTHER READING

1. Instructions for a ghost engine

The Menebrea translation is available online at psychclassics.yorku.ca/Lovelace/menabrea.htm

Thony Christie's analysis of Ada King's programming contribution is on his blog The Renaissance Mathematicus: thonyc.wordpress.com/2012/12/26/christmas-trilogy-2012-part-ii-charles-and-ada-a-tale-of-genius-or-of-exploitation

Grover's quantum search paper 'Quantum Mechanics Helps in Searching for a Needle in a Haystack' is available online at https://arxiv.org/abs/quant-ph/9706033

2. Making a world, bit by bit

General history of computing: *The Universal Machine*, Ian Watson (Copernicus, 2012)

History of computing from a personal computing viewpoint: *When Computing Got Personal*, Matt Nicholson (Matt Publishing, 2014)

Alan Turing: *Turing: Pioneer of the Information Age*, Jack Copeland (OUP, 2014)

John von Neumann: *John von Neumann: The Scientific Genius Who Pioneered the Modern Computer*, Norman Macrae (American Mathematical Society, 2000)

Universal Turing Machine: Alan Turing's paper 'On Computable Numbers, with an Application to the *Entscheidungsproblem*' is available online at www.thocp.net/biographies/papers/turing_oncomputablenumbers_1936.pdf

3. The soft touch

Algorithms applied to real life: *Algorithms to Live By: The Computer Science of Human Decisions*, Brian Christian and Tom Griffiths (William Collins, 2016)

4. Quantum strangeness

General introduction to quantum physics: *Cracking Quantum Physics*, Brian Clegg (Cassell, 2017)

Exploration of quantum physics applications: *The Quantum Age*, Brian Clegg (Icon Books, 2015)

Quantum entanglement: *The God Effect*, Brian Clegg (St Martin's Press, 2006)

EPR paper: Albert Einstein, Boris Podolsky and Nathan Rosen's paper 'Can Quantum-Mechanical Description of Physical Reality Be Considered Complete?' is available online at journals.aps.org/pr/pdf/10.1103/PhysRev.47.777

5. Quantum algorithms

Logic: *The Art of Logic*, Eugenia Cheng (Profile Books, 2018)

Probability and life: *Dice World*, Brian Clegg (Icon Books, 2013)

Cocktail stick dropping simulator: ogden.eu/pi/

Using quantum computers for Monte Carlo methods: Ashley Montanaro, 'Quantum Speedup of Monte Carlo Methods', *Proceedings of the Royal Society A: Mathematical, Physical and Engineering Sciences* 471.2181 (2015), p. 20150301: arxiv.org/pdf/1504.06987.pdf

Quantum computers and option pricing: Nikitas Stamatopoulos et al, 'Option Pricing Using Quantum Computers', *Quantum* 4, 291 (2020): arxiv.org/pdf/1905.02666.pdf

6. Quantum hardware

Feynman's talk on quantum computing: Richard P. Feynman, 'Simulating Physics with Computers', *International Journal of Theoretical Physics* (1982): catonmat.net/ftp/simulating-physics-with-computers-richard-feynman.pdf

Trapping single photons: Michele Cotrufo et al, 'Excitation of Single-Photon Embedded Eigenstates in Coupled Cavity–Atom Systems', *Optica* (2019)

IBM quantum computing in the cloud: www.ibm.com/quantum-computing/technology/experience

Google's quantum supremacy paper: Frank Arute, Kunal Arya, Ryan Babbush et al, 'Quantum Supremacy Using a Programmable Superconducting Processor', *Nature* 574, 505–510 (2019)

D-Wave quantum computers: www.dwavesys.com

7. To infinity and beyond

Artificial intelligence: *Artificial Intelligence: Modern Magic or Dangerous Future?*, Yorick Wilks (Icon Books, 2019)

INDEX

ALSO AVAILABLE

How does big data enable Netflix to forecast a hit, CERN to find the Higgs boson and medics to discover if red wine really is good for you? And how are companies using big data to benefit from smart meters, use advertising that spies on you and develop the gig economy, where workers are managed by the whim of an algorithm? Acclaimed science writer Brian Clegg assesses the good, the bad and the ugly of this fast-developing field.

ISBN 9781785782343 (paperback) / 9781785782497 (ebook)

AI expert Yorick Wilks takes a journey through the history of artificial intelligence up to the present day, examining its origins, controversies and achievements, as well as looking into just how it works. He also considers the future, assessing whether these technologies could menace our way of life, but also how we are all likely to benefit from AI applications in the years to come. Entertaining, enlightening, and keenly argued, this is the essential one-stop guide to the AI debate.

ISBN 9781785785160 (paperback) / 9781785785177 (ebook)